D1611608

EVOLVING THEORIES OF PUBLIC BUDGETING

RESEARCH IN PUBLIC ADMINISTRATION

Series Editor: Jay D. White

Recent Volumes:

Volume 3: Edited by James L. Perry

Volume 4: Edited by Jay D. White

Volume 5: Edited by James L. Perry

RESEARCH IN PUBLIC ADMINISTRATION VOLUME 6

EVOLVING THEORIES OF PUBLIC BUDGETING

EDITED BY

JOHN R. BARTLE

Department of Public Administration, University of Nebraska at Omaha

2001

JAI imprint
An imprint of Elsevier Science

Amsterdam – London – New York – Oxford – Paris – Shannon – Tokyo

ELSEVIER SCIENCE Ltd
The Boulevard, Langford Lane
Kidlington, Oxford OX5 1GB, UK

First edition 2001

Library of Congress Cataloging in Publication Data
A catalog record from the Library of Congress has been applied for.

British Library Cataloguing in Publication Data
A catalogue record from the British Library has been applied for.

ISBN: 0-7623-0790-0

∞ The paper used in this publication meets the requirements of ANSI/NISO Z39.48-1992 (Permanence of Paper).

Printed in The Netherlands.

CONTENTS

LIST OF CONTRIBUTORS

John R. Bartle	Department of Public Administration University of Nebraska at Omaha
John P. Forrester	U.S. General Accounting Office
Mark T. Green	Atkinson Graduate School of Management Willamette University
C. Jeff Hartley Jr.	Department of Political Science The University of Alabama
Paula S. Kearns	Program in Public Policy and Administration, Michigan State University
Jun Ma	Department of Public Administration University of Nebraska at Omaha
Janet Foley Orosz	Department of Public Administration & Urban Studies, University of Akron
John W. Swain	Institute for Public Policy and Administration, College of Business and Public Administration, Governors State University
Fred Thompson	Atkinson Graduate School of Management Willamette University
Kurt Thurmaier	Department of Public Administration University of Kansas
Katherine Willoughby	Andrew Young School of Policy Studies Georgia State University

PREFACE

John R. Bartle has put together an impressive collection of original essays on public budgeting. Each essay is thoughtful and informative, adding to the knowledge base of the field.

As the Series Editor for *Research in Public Administration*, I invite scholars in the field to serve as editors of future volumes on specific topics of importance to the field. One of the advantages of publishing in this research annual is that it can accept papers that journals would not usually publish because of their length. If you wish to discuss a topic for a future volume of *Research in Public Administration*, please contact me at the address below.

Jay D. White
Department of Public Administration
University of Nebraska at Omaha
Omaha, NE 68182
phone: 402-489-6767
email: jdneb@concentric.net

1. SEVEN THEORIES OF PUBLIC BUDGETING

John R. Bartle

SCOPE OF THIS VOLUME

Public budgeting has existed since the first government, in the sense that resources were collected and then used by the sovereign. Thc budgeting process that most Americans are familiar with did not evolve in the U.S. until the early 1900s when the executive budget reform was generally adopted. Budgeting as a conscious field of study in the U.S. followed this development. While budgetary reform was central to changes in the Progressive Era, its focus was prescriptive and process-oriented, leading V. O. Key (1940) to write, "American budgetary literature is singularly arid . . . [T]he absorption of energies in the establishment of the mechanical foundations for budgeting has diverted attention from the basic budgeting problem (on the expenditure side), namely: On what basis shall it be decided to allocate x dollars to activity A instead of activity B?" (p. 1137). In 1979, Robert Inman (1979) wrote, "we have only begun to understand the processes of turning public dollars into public outputs [and] . . . [w]e are still further from understanding the important role of the political choice process in determining budget outcome" (p. 312). In a sense, both of these eminent scholars are correct. There is no single theory of budgeting, and scanning any budgeting text makes this clear. However, most budgeting research is not atheoretical; a variety of theories of policymaking and politics have been applied to the field. There is no single theory, there are several.

Evolving Theories of Public Budgeting, Volume 6, pages 1–9.

ISBN: 0-7623-0790-0

The problem for students of budgeting and for scholars doing research in the area is that there is often little recognition that there is such a diversity of theories. As a result, a study from one theoretical perspective is hard to square with another study from a different perspective. The result is that much research "talks past" other work and it is difficult for scholarship to accumulate. This is in part due to the fact that budgeting has been studied by scholars from a number of different disciplines. As Irene Rubin (1988) observed, "[t]hose trained in only one academic discipline are likely to approach budgeting from the premise and within the theory of their own discipline . . . None of this adds up to meaningful budgetary research agendas. Multidisciplinary work has no cumulative focus" (p. 7).

Some ten years ago, freshly out of my doctoral program and eager to conquer new worlds, I speculated why the study of public budgeting does not have a dominant paradigm in the same way that economics does. It struck me that the term "budget theory" was as unfamiliar as the terms "microeconomic theory" or "price theory" were familiar. Shortly afterwards, I was surprised and excited to see a panel on budget theory at the annual conference of the Association for Budgeting and Financial Management (ABFM). Since then, one of the questions I have pursued has been: which theories of budgeting are useful? Ultimately, that pursuit has led to this book.

In editing this book, I asked the authors to try to answer this question: "If a doctoral student came into your office and said 'I want to do research in public budgeting from *this* perspective,' what would you tell them? What should they read, what are the interesting questions, and what is the long-run potential of the theory?" I have tried to let the voices of the authors speak with as little interference as possible, and so the chapters follow different styles, formats, and theoretical values. I believe it was important not to impose a set of criteria on the authors as to what the characteristics of a "good" theory are. Instead they present their own case, either for or against that theory. The reader is free to judge the merits of these arguments. With seven intriguing theories to compare, I believe this will clarify the reader's thinking and theoretical values.

This is not a mystery thriller. The reader will probably not find that one of these theories is the heir apparent to the title of the new theoretical paradigm for public budgeting research. Instead they will probably find that all of these theories have important strengths and weaknesses, and that some are more appealing to them than others. This is healthy. I believe that there are many reasons why there is no dominant paradigm in public budgeting research. As a relatively young and multi-disciplinary field of study, it has been subject to influences from various authors. Much research in budgeting is practice-oriented, rather than theoretical. And there are differences of opinion on epistemological issues. Theory in public budgeting is *evolving*, not static, which makes it a very intellectually exciting field of study. My hope for this book is that it stimulates future researchers to see the opportunities in this area, and to improve upon these and other theories. Then people may not snicker when someone refers to themselves as a "budget theorist."

This book puts together seven critical essays on different theories that have been applied to public budgeting. The theories examined are: incrementalism, a budget process model, the organizational process model, the median voter model, public choice theory, a post-modern perspective on budgeting, and a transaction cost model. There are other theories that might have been included in this book. Their exclusion is not calculated; rather they are due to my limitations as an editor. However many of the important perspectives that have been brought to bear on public budgeting are here. Hopefully future collections will include other theories and update those covered here. But I do believe that some very important theories are discussed here in a real depth that will help many researchers. More than review articles, these chapters have numerous critical insights and original developments, some of which have exciting potential. Hopefully they will be productive soil for many flowering dissertations.

REVIEW OF THE CONTENTS

Incrementalism

Incrementalism is probably the most widely-known theory of budgeting. In Chapter 2, John Swain and Jeff Hartley review this theory, as articulated by Aaron Wildavsky (1964). They also review some of the literature elaborating and testing the model, and summarize the criticism of the model. They break Wildavsky's model of incrementalism into four elements that serve as their basis for evaluating the utility of the theory. These elements are: a qualitative descriptive argument ("what is"), empirical theory ("explains what is"), analytical theory ("what must be"), and normative theory ("what should be"). With regard to each of these elements of incrementalism, respectively, Swain and Hartley find that:

- Incrementalism is a superior theory for describing the process of public budgeting.
- Empirical findings fit observed behavior better than any other budget theory.
- It provides a "realistic framework" for analyzing strategic decisions about budgeting, and deducing expected behavior.
- Incrementalism is a "reasonable and credible perspective" that is normatively

defensible because it reflects political power as opposed to rationalist perspectives on budgeting which may be unrealistic and undemocratic.

They conclude that incrementalism provides a unifying perspective and a common vocabulary that can and should serve as the basis for future research. The criticism of the theory provides research questions, such as that advanced recently by Jones, True and Baumgartner (1997).

The Multiple Rationalities Model of Budgeting

The process of budgeting, and indeed policy-making in general, has been a central element of study in public administration. The most widely used model is the policy process model, also known as the "stages heuristic" model. This traces the policy process through agenda setting, policy formulation and legitimation, policy implementation, and policy evaluation. In Chapter 3, Kurt Thurmaier and Katherine Willoughby discuss models of budgeting that use this approach, but they take a bold and original step by linking previous research on individual decision making to system-wide theories of the policy process and budgeting specifically. This provides a crucial link that enriches and elaborates these theories in a much needed way. Until now, the unit of analysis of process models had not been extended to the individual level in a rigorous way.

They argue that the central budget bureau is the vortex into which policy and budget decisions are drawn. Budget bureau analysts learn to understand agendas, judge political feasibility, interpret signals, and sense timing in order to align micro-budgeting decisions by agencies, Governors, and appropriators with macro-budgeting constraints and objectives. When a window of opportunity opens, analysts can advocate or help advocate on behalf of an agency. In this way, they argue, budget analysts are critical links between micro-budgeting opportunities and macro-budgeting considerations. This then helps explain how budget proposals move through the policy process, and how individual decision-making styles influence outcomes. Further, what is feasible in this context is situationally-determined. A skilled actor can influence the agenda by correctly judging political feasibility and being ready with specific, relevant proposals. Thus, their model takes on an interpretive flair.

The Organization Process Model of Budgeting

In Chapter 4, Mark Green and Fred Thompson review what they term "organizational process models of budgeting." These models focus on the decision as the unit of analysis, and employ bounded rationality as the behavioral assumption. John Crecine, Patrick Larkey, and John Padgett are some of the leading authors in this area. This theory models budgetary decisions by agents using decision routines, often modeled using binary logic. The decisions of any agent then fit within a web of decisions by other agents, ultimately creating a system that can be simulated. These simulations can either be used to predict and explain decisions, or to estimate counter-factual behavior. For example Larkey simulated city spending in the absence of the U.S. federal general revenue sharing program in order to determine the change induced by the program (Larkey, 1979).

Green and Thompson compare the organizational process model to the rational choice model of public budgeting and finance (essentially the median voter model, discussed in Chapter 5). They find the organizational process model to be a worthy alternative deserving of further elaboration. While the organizational process vein of literature is small and not presently attracting substantial attention, they believe it is a rich and potentially productive vein to mine.

The Median Voter Model

The median voter model is an approach widely used by economists examining public expenditure decisions. It analyzes government fiscal choice using the economic model of consumer choice. It is assumed that a single individual – the median voter – is decisive in local budgetary decisions and the resulting public budget is then a reflection of his or her preferences for public goods. In Chapter 5, Kearns and Bartle examine the development of this model and evaluate its usefulness for future research in budgeting and expenditure research. They argue that the model is parsimonious, has high explanatory power, and can be used to estimate important variables such as price and income elasticities. However its deficiencies are serious:

- It abstracts from the influences of institutions, the distribution of power, and the process of consideration.
- It does not perform well compared to other models in rigorous specification tests.
- It does not perform well in modeling the fiscal behavior of representative, general purpose local governments, or state governments.
- It presents some anomalies, such as the famous "flypaper effect," and
- It does not include a model of the supply side of public services.

Kearns and Bartle conclude that future prospects for the median voter model as a vital theory of public budgeting are pessimistic.

Public Choice and Public Budgeting

In Chapter 6, John Forrester contrasts public choice theory with public administration generally, and the literature on public budgeting in particular. Earlier comparisons and debates between these approaches have often been heated (Golembiewski, 1977). In a careful, even-handed way, Forrester examines similarities and differences between the two perspectives. Public choice theory has been more doctrinaire, yet it makes explicit assumptions and tests them rigorously. Public budgeting research generally does not have a central paradigm, and is more pragmatically driven. But both approaches study government choices, have similar views about bureaucracy, focus on fiscal changes, focus on budgetary agents, and normatively accept an economizing orientation. As a result, Forrester sees a potential for budgeting research to productively draw from public choice theory, if for no other reason that changes in government (such as tax limitations, school choice, and federal spending cuts) advocated by public choice theorists are affecting government budgetary processes and outcomes.

Post-Modern Budgeting

In other areas of public administration, a theory of growing importance is post-modernism, which is a broad umbrella used to refer to interpretive, constructivist, subjective, and critical theories of administration. In Chapter 7, Janet Foley Orosz argues that, because budgeting takes place in the social world, it can and should draw from the subjective theories of organization. Budget actors engage in rituals, use language, and manipulate symbols to create meaning. They may also do so to manipulate the budget and other levers of power for good or ill. A shared understanding allows actors to interpret relevant events and data in order to justify a course of action. Especially in an uncertain environment of fiscal stress where the accepted view of reality is brought into question, finding a new meaning is required for a collective organization to take action. Ultimately, a budget becomes a negotiated reality where both meaning and value are intersubjectively created by the participants. This might help explain the challenge of meaningful citizen participation in budgeting, as until citizens and administrators understand and accept the meaning that others associate with the budget and its policies, a meaningful dialogue cannot take place.

Orosz identifies seven local narratives of public administration, and draws parallels from them to eras and traditions in budgeting. These are: constitutionalism, the politics-administration dichotomy, scientific study and practice of public administration, the interplay between theory and practice, the "new public administration" of the Minnowbrook conference, feminist theory, and the "new public management" of the 1990s. While unorthodox for many researchers in budgeting, the approach of these theoretical perspectives deserves their attention. Interpretive theories have had a very important influence recently on public management, and certainly budgeting is an aspect of management. Successful applications of these sorts of models is likely to be a major growth area in budgeting research.

Transaction Cost Theory

In Chapter 8, Bartle and Ma look at an emerging viewpoint from the new institutional economics – transaction cost theory. This perspective applies a broad model of cost-minimization to study the creation and role of institutions. Transaction costs in any exchange are attributable to asset specificity, informational asymmetry, organizational rules, and customs. The presence of these factors will affect the size and role of organizations. This perspective has mainly been used to examine the structure of firms in private markets, although increasingly it has been applied to political exchanges. Its application to the study of public budgeting and financial management has been very limited, despite what seems like a natural fit.

Bartle and Ma review the relevant literature, and assert that:

- Political exchanges in budgeting fit closely the logic of the theory,
- The theory's focus on institutions, comparative social science, and historical development fit well with traditions of public administration, and that
- Its generality allows it to be the organizing rubric for a variety of disparate studies.

They believe it could be productively applied to the study of many financial management arrangements in government, such as purchasing, contracting, cash management, debt management, and other housekeeping functions. It could also be applied to studying traditional budgeting subjects, such as appropriation rules, budget reforms, and accounting fund structure. However as an economic model rooted in the values of economic efficiency, it neglects political influences. As this theory is very new, it is far too early to pass judgment, however this approach may be useful for future research in public budgeting and financial management.

A COMMENT

Although some would disagree, I believe that public administration is a field of study, not a discipline. The same is true of public budgeting. As such it is at the intersection of many disciplinary roads and is naturally prone to operating simultaneously with different theoretical perspectives in use. While this may confuse those who come to the field looking for theoretical coherence, for those who can be comfortable with some degree of ambiguity, it is a rich, intellectually exciting area to study.

At the same time, problems come when one tries to understand different studies from different theoretical perspectives. It can also be overwhelming for the researcher beginning their inquiry: what theoretical perspective should one choose? Or perhaps more pragmatically, should one begin with a theory and then go gather data, or does one collect data first, and then attempt to piece together a theory? My answer would be: either. There is plenty of good research from either side of these approaches, and ultimately the proof is in the pudding. The real problem comes when different theoretical perspectives are not compared, and our knowledge does not accumulate very well. This volume is an attempt to address this in a small way. Most of the authors in this volume find at least some, if not a great deal of merit in the perspective they examine. Other meritorious perspectives omitted from this volume could certainly be added. This is good news, because it tells us that the field is alive and well with a rich mixture of competing ideas. Theories are regularly being modified, extended and re-examined.

Some, perhaps much, of the literature on budgeting is not very consciously theoretical. It is more focused on description and technique. However these are still the building blocks of theories and can often prove helpful. As a field of study that strives for *relevance*, a constant interaction with the field – our laboratory – is helpful if not essential. Perhaps Tucker (1982) puts it best,

> Past findings of budgetary research raise more questions than they answer; some of the most provocative questions have been raised in the spirit of challenging conventional wisdom. Rather than forbidding future inquiry into 'refuted' theories, we should encourage theoretical pluralism in budgeting research. The state of our knowledge justifies neither adherence to nor proscription of any single approach to modeling the budgetary process (p. 336).

ACKNOWLEDGMENTS

I have many debts in this project. My colleague Jay D. White gave me the opportunity to organize this volume, despite his never having taken a course in budgeting. He, and Elsevier Publishing demonstrated great patience in awaiting the emergence of this volume. I am lucky to be able to call all the chapter authors friends of mine whose work I have great respect for. I also owe them many thanks for their patience, and their willingness to share their significant work in this forum. The reviewers of this volume demonstrated great knowledge, effort, and service by taking the time to review these chapters. They are: William Duncombe of Syracuse University, Rebecca Hendrick of the University of Illinois at Chicago, and Richard Box, Carol Ebdon, and Dale Krane of the University of Nebraska at Omaha. Joyce Carson of the Center for Public Affairs Research at the University of Nebraska at Omaha gave of her time to edit these chapters with a professionalism that I greatly admire. I also appreciate the numerous ways my university supported me in this project. Finally, I owe a priceless debt to my wife and children for their understanding.

REFERENCES

Golembiewski, R. T. (1977). *Public Administration as a Developing Discipline, Part 1: Perspectives on Past and Present.* New York: Marcel Dekker Inc.

Inman, R. P. (1979). The fiscal performance of local governments: An interpretive review. In: P. Mieszkowski & M. Straszheim (Eds), *Current Issues In Urban Economics* (pp. 270–321). Baltimore: The Johns Hopkins University Press.

Jones, B. D., True J. L., & Baumgartner, F. R. (1997). Does incrementalism stem from political consensus or from institutional gridlock? *American Journal of Political Science, 41*, 1319–1339.

Key, V. O., Jr. (1940). The lack of a budgetary theory. *American Political Science Review, 34*, 1137–1140.

Larkey, P. D. (1979). *Evaluating public programs: The impact of general revenue sharing on municipal government.* Princeton NJ: Princeton University Press.

Rubin, I. S. (1988). *New directions in budget theory.* Albany NY: State University of New York Press.

Tucker, H. J. (1982). Incremental budgeting: Myth or model? *Western Political Quarterly, 35*, 327–338.

Wildavsky, A. (1964). *The politics of the budgetary process.* Boston: Little, Brown and Company.

2. INCREMENTALISM: OLD BUT GOOD?

John W. Swain and C. Jeff Hartley, Jr.

ABSTRACT

Incrementalism is certainly old, but is it still good? Incrementalism is examined here through reviewing key pieces of literature from that perspective, both the seminal works by Aaron Wildavsky and the later work he did with Otto Davis and Michael Dempster. The essential features of incrementalism are identified, criticisms of incrementalism are laid out, and an assessment of incrementalism is made. The overall assessment is that the criticisms have some validity, but they do not add up to a refutation or discrediting of incrementalism. Incrementalism provides a basis for developing budget theory, especially positive budget theory.

INTRODUCTION

The incremental theory of budgeting (hereafter incrementalism) is well-known, having been evangelized by an outstanding political scientist who is the most notable author on public budgeting: Aaron Wildavsky. Incrementalism is also descriptively elusive, normatively controversial, and empirically difficult to test. It abstractly states characteristics of, and relationships in, public budgeting. Describing it adequately requires using unambiguous terms. Discussing it normatively requires engaging in evaluating diverse values and outcomes. Testing it requires operationalizing it in empirical terms, gathering data, and

Evolving Theories of Public Budgeting, Volume 6, pages 11–27.

ISBN: 0-7623-0790-0

appropriately analyzing that data. Because of all of these features, examining incrementalism is a difficult task that we pursue by looking at it with fresh eyes, open minds, and a critical attitude.

The question to answer here is how incrementalism can contribute to future research in public budgeting. Before addressing that question, we survey and critically assess the major incrementalism literature and criticisms of that perspective.

LITERATURE SURVEY

Incrementalism Presented

The Basic View

Aaron Wildavsky's *The Politics of the Budgetary Process* (1964) is the best-known expression of incrementalism. Three revised editions of that book were published in 1974, 1979 and 1984. A similar book with a slightly different title, *The New Politics of the Budgetary Process*, was published in 1988, and subsequent editions were published in 1992 and 1997. Despite massive changes in budgeting, incrementalism has endured essentially unchanged over the thirty-year span of the two volumes.

The best starting points for understanding incrementalism as expressed by Wildavsky are to examine: (1) the context of budgeting literature and practice at the time when the book was first written; (2) a related earlier piece of writing, and; (3) the preface of the original book.

The contexts of budgeting literature and practice diverged when Wildavsky first wrote his book. With few exceptions, budgeting literature since early in this century had been unequivocally reformist (Wildavsky, 1964, 127–128; Burkhead, 1956, 12–29). Public budgeting reform efforts emanating from the Progressive movement and its heirs were oriented toward greater degrees of centralization, technical sophistication, and comprehensiveness. The promised blessings of budgetary reforms, made possible by the application of technical expertise and wise decision making, included greater accountability, economy, efficiency, and effectiveness, among other good things (Burkhead, 1956, 9–29 and 306–339; Schick, 1966). As practiced, budgeting displayed two sides. Technically, detailed line items and efforts to ensure accountability were noticeable. Politically, legislatures dominated.

Wildavsky outlined the argument of *The Politics of the Budgetary Process* in an article titled "Political Implications of Budgetary Reform" (1961). That article starts by directly responding to reformers, specifically addressing V. O. Key's basic budgeting question ("On what basis shall it be decided to allocate

X dollars to Activity A instead of Activity B?") and subsequent attempts to answer it. Wildavsky argues that the question as stated is unanswerable without the acceptance of a comprehensive and specific normative theory of politics, which he views as impossible. Wildavsky points out that budgeting decisions in the United States are made through the political process, which reformers criticize as being irrational. Changes in budgeting mean changes in politics and in outcomes. Wildavsky notes that very little is known about actual budgeting and sketches out many ideas later found in his book. Wildavsky suggests that developing a more adequate theory of budgeting would provide a little knowledge, that a little knowledge should proceed a lot of reform, and that a lot of reform might produce much in the way of unintended consequences. Incrementalism, though recognizable, is much less developed here than in the later book.

The preface of *The Politics of the Budgetary Process* (1964) supplies explanations, unlike any later account, of how Wildavsky went about researching and writing that book. First, he pursues the underlying research question, "How does the budget (the funds appropriated by Congress and actually spent) get made in American national government?" (v). Second, he researches the book by interviewing participants in that budgeting process ("agency heads, budget officers, Budget Bureau staff, appropriations committee staff, and Congressmen") and by reading the appropriations hearings for twenty-five agencies over a fifteen-year period (v). He asks how they decided budgetary amounts and how they tried to have their decisions ratified in the budgetary process. Third, he states his research task as describing, explaining, and appraising the process. Fourth, he notes that the work is limited in scope and is a point of departure rather than anything else: "No one is likely to mistake this small volume for a comprehensive or definitive work on the subject of budgeting" (vii). Significantly, despite universalistic rhetoric, the work explicitly concerns one level of American government at one period of time.

The incremental perspective found in *The Politics of the Budgetary Process* (1964) is derivative from the work of Charles Lindblom (Wildavsky, 1961, 183, 185; Wildavsky, 1964, 131). Wildavsky applies Lindblom's ideas to budgeting in a distinctive fashion, however.

In Chapter 1, "Budgets," Wildavsky discusses various understandings of budgets and budgeting that lead up to his own conception that identifies the scope of his small volume. He puts budgeting into the context of political institutions when he offers his conception of "the budgetary process as a phenomena of human behavior in a governmental setting" (4). This conceptual formulation focuses attention especially on the legislative process and tends to preclude the consideration of many factors, e.g. ultimate outcomes of budget decisions and behavior outside of governmental settings. The concluding sentence displays the book's focus clearly when it says that "the budget lies at the heart of the political process" (5).

In Chapter 2, "Calculations," we find that budgeteers are faced with problems of deciding expenditure amounts to request and to spend (agency), to recommend (Bureau of the Budget), and to give (Congress) in the context of pressures from their collective political environment. The political environment of budgeting, which is taken as given, is posited as setting the conditions that effectively limit the choices available to participants in the budgetary process. Those conditions include the U.S. Constitution, group pressures, positions of political parties, the contemporary climate of opinion, and common expectations of how the budget process works (6–7).

In that decision-making context, participants make calculations, i.e. decide what to do, by taking various factors into account. The factors include the phenomena in the political environment and the technical subjects of budgets, e.g. medical research, highways, and national defense. The calculation situation is exceedingly complex because of the aforementioned political environment and the technical subjects and because of the human limitations of time, ability, and knowledge, especially in regard to the need to compare unlike things for the sake of making funding decisions, e.g. recreation relative to education.

Participants use aids in making calculations to simplify deciding: aids are ways of considering some, rather than all, possible alternatives and using some, rather than all, possibly available information in deciding. Aids encompass generic devices used in a wide variety of settings and specific devices used in the budgetary process. Generic devices include using indices, deciding based on experience, satisficing (taking the first minimally satisfactory alternative for a decision), and choosing based on a relatively limited range of meaningful changes from current practices (6–16, 125). The specific devices include the related concepts of fair share and base and the division of labor in the budget process into a system of roles with complementary perspectives. The system of roles has the effect of reducing decision-making requirements by dividing decision-making labors into parts. The two concepts of base and fair share guide decision makers toward assuming that most of an organization's budget is not worth bothering with and that changes in expenditure authority will be widely shared among agencies on some common basis and then behaving accordingly. The two concepts reflect assumptions that thereby become realistic descriptions. A system of roles and perspectives simplifies by focusing the behavior of different budgetary participants within limited bounds according to their institutional location and by providing participants with views on other participants' roles and behaviors. For example, agency personnel advocate expenditures for their agency and see themselves as serving the public. In this chapter on calculations, Wildavsky shows budgeteers responding to human limitations and difficulties in the face of complexity largely by simplifying decisions rather than by increasing their calculational effort.

In Chapter 3, "Strategies," Wildavsky discusses strategies used by agencies to secure funding as a part of the calculational process, although the behavior of other budgetary participants is also taken into account. Although technical knowledge of the subject matter being budgeted for and the associated estimates are said to be desirable, technical knowledge is denigrated in favor of being a "good politician" (64). Three categories of strategies are associated with being a good politician. The two ubiquitous types of strategies involve cultivating an active clientele and developing confidence among other governmental actors. The first reflects the need for popular support in a system of popular government. The second reflects the need for committee members to rely on agency officials because members and their staff are very limited in what they can observe. The more trust an agency engenders, the better it will be treated. Confidence strategies include developing personal trust and trust in the actual achievements of the agencies, i.e. results of programs. The third type of strategies are contingent strategies designed to shape interpretations of events so that agencies can defend their budget bases in the sense of maintaining appropriations levels, increase their budget bases in the sense of gaining more appropriations, and expand their budget bases in the sense of gaining new program. The chapter concludes with a discussion of the topic of empirically explaining and predicting budget outcomes (123–126).

In Chapter 4, "Reforms," budget reforms are discussed as involving the choice between traditional budgeting, which means the patterns of behavior discussed by Wildavsky, and some more rational form of budgeting as advocated by reformers. Traditional budgeting is political budgeting that reduces the amount of calculation and facilitates agreement. Reform proposals would increase calculational burdens, inhibit agreement, and change the results of the budgetary process in the sense of budget shares devoted to different programs by altering the political process. Budgeting reform, Wildavsky asserts, is not politically neutral in respect to outcomes. Otherwise, why bother with reform? Proposed reforms would necessarily include reforming the political system as a means of reforming budget outcomes.

In Chapter 5, "Appraisals," Wildavsky weighs the advantages and disadvantages of both traditional and rational budgeting and concludes that traditional budgeting is clearly superior. Wildavsky addresses criticisms of traditional budgeting and endeavors to show that the purported shortcomings of traditional budgeting provide advantages that make it superior to rational budgeting.

Although not claiming to have found the only right perspective, we offer the following basic conception of incrementalism based on *The Politics of the Budgetary Process*. *Basic* here refers to the essential features of incrementalism, its fundamental tenets so to speak. Wildavsky presents points that generally characterize budgeting and points that are particular to some actual place and time. The interpretative problem seems to be separating the common from the specific and the descriptive from the requisite because each would seem to require different standards of proof. First, in contrast to the reform view of budgeting as technical problem solving, budget decisions are political because they require approval of governmental authority to spend money. The political environment provides inputs for the decision-making process. The political environment includes public opinion and support for particular agencies' proposed expenditures (Wildavsky, 1964, 4–7).

Second, the outcomes of budgetary processes (appropriations) are generally incremental for a government and its major organizational units. The outcomes vary relatively little from one budget cycle to another and from one stage of a budget process to another. More than three-quarters of observations of thirty-seven domestic agencies over a twelve-year period showed a change of less than 30%, and those above that level could often be explained as involving cyclical activity, e.g. the Census Bureau (Wildavsky, 1964, 14). Wildavsky characterizes major shifts in expenditures, which are exceptions, as "shocks" to budgeting systems (126).

Third, budgeting can be understood as decision making where officials use calculations that simplify because of complexity. *Calculations* means deciding what to do; *simplification* refers to using some rather than all alternatives and some rather than all possibly available information in deciding. In addition to generic devices, specific devices include the concepts of base and fair share and a system of roles and perspectives that divide the labor of decision making.

Fourth, within specific roles, participants engage in particular strategies, which are behaviors designed to achieve goals that are based on calculations. Wildavsky argues that participants, on the basis of experience, choose strategies that they believe are most likely to succeed under conditions of uncertainty (62–64).

Specific details of the federal budgetary process are peripheral to the basic conception of incrementalism because that process is only one case. Incrementalism should not be judged based on one case, interesting as it may be.

The Politics of the Budgetary Process gradually glides from being a tentatively framed descriptive work on one of many possible conceptions of budgeting in the first chapter to expressions of relationships in budgeting and limitations on human decision-making capacity in the second and third chapters to more of a normative focus in the two concluding chapters. Wildavsky weaves together four different kinds of arguments: qualitative descriptive, empirical theory, analytical theory, and normative theory. The different kinds of arguments can be thought of in terms of different kinds of verbs applied to the "whats" of budgeting: what is, explains what is, what must be, and what should or ought to be. The qualitative descriptive arguments concern observable behavior: these qualities are evident in behaviors occurring in the budgeting process. The empirical theory arguments, although tentative, suggest explanations for behavior and hold out the promise of predicting outcomes: this observable behavior may be explained and perhaps predicted by these factors. The analytical theory arguments posit assumptions about the budgetary process and logically derive consequences of those assumptions: because these conditions exist, these behaviors occur and others do not. The normative theory arguments relate features of the traditional budgetary process, and proposed reforms, with American political norms: these features are beneficial or not relative to these norms. Wildavsky integrates the four kinds of arguments into an holistic conception of budgeting. The qualitative descriptive arguments presented above suffice for present purposes. We discuss the analytical theory and normative arguments here briefly, but we focus more on empirical theory arguments because we believe that they hold the most promise for advancing budgetary theory.

Analytical Theory and Normative Theory

Incrementalism may be expressed in terms of relationships between human characteristics (analytical theory) and in terms of manifesting particular values (normative theory). As analytical theory, incrementalism provides a logical construct of budgetary processes. Assumptions are made, the logic is specified, and details of the process are deduced. For example, the purposes and roles of budgetary actors lead to calculations and strategies. The deductive process leads to conclusions about why the budget process proceeds the way it does as the meaning of the assumptions becomes apparent. For example, human limitations in regard to calculations make merit mean something other than the sum of calculations for all possible values; instead, merit is judged by political support (Wildavsky, 1964, 176–177).

Normatively, incrementalism is associated with advancing particular values, including popular government, political participation, the minimization of conflict, and realistic (limited) expectations of budgeting (Wildavsky, 1964, 145–180). Simply put, for Wildavsky budgeting is politics. Reformers' desires to change budgeting to change budgetary outcomes, whether because of ignorance of what is involved (1961, 185) or because of dislike of politics

(1966, 308), necessarily require changes in the political system. Wildavsky argues normatively for the values of the current political system and political decision making generally against the argument that budgeting would be and should be better by making it more rational and thereby produce greater efficiency.

Empirical Theory

Subsequent to writing and publishing *The Politics of the Budgetary Process* and a voluminous amount of other works, Wildavsky wrote on empirical budget theory. Specifically, he wrote on quantitative explanatory and predictive models of incrementalism with Otto Davis and Michael Dempster, focusing on the U.S. federal government.

The most frequently cited empirical piece establishes that presidential budget requests and congressional appropriations for fifty-six federal agencies between 1947 and 1963 can be described generally as being "similar to those produced by a set of simple decision rules that are linear and temporarily stable" (Davis, Dempster & Wildavsky, 1966a, 529). Budget requests are best explained by prior year appropriations, and appropriations are best explained by budget requests. The equations representing those relationships posit the resultants as a function of an initial value plus or minus a random value. The correlation coefficients (adjusted) for most observations exceed 0.9. Two sorts of exceptions to the generally stable pattern were found. First, agency budget requests or appropriations occasionally deviate from the pattern temporarily and return to the previous pattern. Such deviations can be explained by specific "exceptional circumstances" (541). Second, some observations constitute shift points, similar to shocks mentioned above, that signify the movement of an agency's budget figures from one level or pattern to another. Shift points often seem to be associated with political events such as changes in presidential administrations and congressional majorities. The discovery of shift points puts predictive models beyond the reach of the authors in this piece. As described later, shift points manifest themselves abruptly in budgetary figures but represent a displacement resulting from a build up of pressure over time (Dempster & Wildavsky, 1982, 278). A related article discusses the selection of the linear equations and the data analysis (Davis, Dempster & Wildavsky, 1966b).

In 1971, partially in response to criticisms of the 1966 work, the authors report corrected statistical procedures, expand the scope to all non-defense agencies for the same time period, and repeat the analysis. The results provide even stronger support for their conclusion that the federal budgeting process behaves "equivalent to a set of temporally stable linear decision rules" (Davis, Dempster & Wildavsky, 1971, 320). They categorize factors for shift points for the periods studied as partisan controversy, new law, external variable, accounting, congressional supervision, reorganization, appropriations policy, and ones not yet identified. They suggest looking at shift points in predicting budgeting outcomes.

Using the same agencies for the same years as in the 1966 work, the authors develop predictive models for fifty-three agencies (Davis, Dempster & Wildavsky, 1974). In addition to the linear equations of the earlier work, they introduce political, administrative, economic, and social variables. The expanded equations not only improve the explanation of budget requests and appropriations but also prove superior to all other quantitative methods in predicting the same agencies' budget figures for 1964–1968 budget years (447).

Dempster and Wildavsky (1982) summarize and extend their work with predictions of executive branch budget requests by finding essentially that agency requests are divided into ones exempt from cuts and ones subject to proportional (fair share) cuts by the departmental and central budget offices, which increases the accuracy of their predictions on those values (288). Also, they offer an interesting econometric model of the relationship between the federal budget and the economy. A later version of this work displays fewer details and a slightly different emphasis on predicting appropriations but not the econometric model (Dempster & Wildavsky, 1986, 81). Empirical studies of incrementalism conducted by other authors mostly support incrementalism (Hoole, Job & Tucker, 1976; LeLoup, 1978, 492–493; Padgett, 1980, 354; Olson, 1987, 57–64).

Incrementalism Criticized

Incrementalism overcame conventional wisdom on budgeting to become the dominating conventional wisdom, which thereby made it a target for critics. An extremely brief overview of criticisms follows. First, normative arguments posit that incrementalism is too hostile to rational decision making and too supportive of the then-existing pluralistic political process (Schick, 1966, 1969). Wildavsky is said to assume that people are satisfied with the outcomes of budgeting and that they have little hope for improved budget decision making. With other pluralists, he assumes the beneficence of the process without examining the actual outcomes. If those conditions change so that people are dissatisfied with outcomes and hopeful of better budgeting, then budget reforms are normatively preferred. Also, Schick notes a paradoxical aspect of Wildavsky's incrementalism when he indicates that Wildavsky's budgetary process is modelled on economists' model of competition that they acknowledge as defective and for which they suggest governmental actions to correct the defects.

The paradox is that the political process presented by Wildavsky has the same kinds of defects and, therefore, is unlikely to supply corrective action (1969, 144–149).

Second, incrementalism is said to describe a historical period (Schick, 1983, 1994; Rubin, 1989). This view generally equates incrementalism with steady, across-the-board increases in yearly expenditures and denies the continued existence of this pattern because of the rise of entitlements, deficits, and changing patterns in federal budget behavior.

Third, incrementalism is said to be confusing and confused in the sense that it can mean many different things but still cannot be pinned down to any particular specific meaning. In this view, multiple meanings, the duality of process and outcome in incrementalism, and the lack of specification of how much an outcome can vary from its preceding value render incrementalism unacceptably vague. Some authors even specify how much an outcome can vary and still be incremental (Bailey & O'Connor, 1975; LeLoup, 1978; Berry, 1990; Gist, 1974).

Fourth, a variety of arguments are made about flaws of the empirical work, specifically concerning the level of analysis or aggregation, definition of incremental, and methodology. Critics argue that analyzing at the agency level yields greater evidence of incrementalism than at less aggregated budgetary figure levels, e.g. programs, and that problem solving proceeds in a more rational manner than the incremental authors suggest (Gist, 1974; Natchez & Bupp, 1973; Tucker, 1982; Williamson, 1967). In 1979, Dempster and Wildavsky responded to critics with "On Change: Or, There Is No Magic Size For an Increment"; they argue that they were specifically examining organizations and budgetary calculations and were not examining policies or problem solving. Also, they introduce a new meaning for incrementalism, i.e. a regular relationship rather than any particular size of increment. Methodological criticisms focus on specific statistical procedures, specifications of the budget process, and the use of longitudinal analysis (Wanat, 1974; Tucker, 1982; LeLoup, 1978; Padgett, 1980; Gist, 1982).

ASSESSMENTS

Despite accepting the validity of many specific criticisms and after carefully reading all of the critical work we could find, we are not convinced that incrementalism has been refuted or effectively supplanted. The most that can be said is that incrementalism has shortcomings.

Common tendencies among those expressing the view that incrementalism is no longer relevant are to dismiss incrementalism in a ritualistic manner without carefully considering it or how it may have informed the critics' own work, to equate criticisms of the empirical work with successful refutation of all aspects of incrementalism, and to misinterpret it to make criticisms of it appear more valid. Examples of these tendencies appear in otherwise generally admirable pieces of work by Meyers (1994) and Stanford (1992). In both cases, empirical criticisms of incrementalism are cited so that incrementalism can be discarded before the authors go on to show a great deal of understanding of budget strategies. Also, Meyers incorrectly presents incrementalism as expressing the view that "budget actors make simple decisions" (6) rather than as involving a desire for "simplification" or decisions being "simplified" (Wildavsky, 1964, 9, 11). It seems to us that budget strategies are primarily interesting within the context of a basically incremental view of budgeting. Also, when citing authors critical of the empirical work, critics often overlook the degree to which the authors they cite agree with or support the incrementalism perspective (Wanat, 1974, 1221; Natchez & Bupp, 1973, 955–956; LeLoup, 1978, 501; Padgett, 1980, 370–371). More than anything else, the currently fashionable style in budgeting literature seems to require rejecting incrementalism while taking an essentially incremental perspective.

Although sorting out criticisms of incrementalism is an intellectually attractive enterprise, our present task requires that we consider how incrementalism can contribute to future research in public budgeting. In doing that, we briefly assess incrementalism in respect to its four different kinds of arguments, with greatest emphasis on empirical theory.

In respect to qualitative descriptive arguments, Wildavsky supplemented incrementalism in the later editions of *The Politics of the Budgetary Process* through its fourth edition initially and then into the first two editions of *The New Politics of the Budgetary Process* with accounts of new phenomena as aspects of the federal budgetary process changed. In the same spirit Naomi Caiden added material to the third edition. Those new phenomena included attempts at implementing new budget approaches and other reforms oriented toward making budgeting more rational, shifts in politics toward greater conflict and concern for budget deficits, and a gradual shift in budgeting away from funding organizations providing services toward funding and influencing activities carried out by other entities, particularly through entitlements and efforts to regulate the economy (Schick, 1994; White, 1994). Also, incrementalism initially did not adequately address the pursuit of budgetary largess by individuals and groups in society because it did not explicitly take into account interest group activity seeking advantage directly through the executive and the legislative branches without agencies seeking funding for their activities. Other important differences in budgeting since the first appearance of incrementalism seem to be the openness of the budgetary process, which decreases reliance on confidence strategies, and the greater degree of conflict among participants in the process, including presidents. Despite these changes, which in themselves are profound, incrementalism still seems to us to describe much of federal budgeting reasonably accurately and also to describe budgeting at other levels of government at least as well as, if not better than, other theories. At the very least, the incremental description of budgeting provides a description of the political process that is clearly superior to reform descriptions of a process that never was and never will be. As far as we are aware, although we have not exhaustively examined this proposition, no other theory provides superior qualitative descriptions of public budgeting in respect to its political dimensions.

In respect to analytical theory, we believe that incrementalism provides a realistic framework for thinking abstractly about budgeting and that it contributes to gaining insights. Limited human capacity to calculate, the desire of agencies to seek funding, the need for political support, and the need to make budgetary decisions create a political environment in which strategies make sense. In contrast, a rational decision-making perspective on budgeting offers little hope of deducing insights. Incrementalism offers a basic framework for analytical theory by identifying conditions that budgeteers face in budgeting and by deducing likely patterns of behavior. Also, we see most treatments of budget strategies as an extension of incremental analytical theory that include specific additional information.

In respect to normative theory, incrementalism seems to offer a reasonable and credible perspective that we believe is at least arguably correct. Its greatest value normatively lies in its use in defending the status quo in budgeting against calls for change by noting the linkage between politics and budgeting and the need to consider how outcomes may be affected by reforms. Wildavsky's challenge to would-be reformers, what difference in outcomes their budget reforms will achieve in politics as well as in budgeting, remains effective (Wildavsky, 1961, 185; Wildavsky, 1964, 127). Although disagreement with the normative perspective is also reasonable, the appropriate starting point seems to be addressing the values expressed by Wildavsky in defense of incremental budgeting rather than assuming that budget reform will produce better results magically. We believe that Schick's 1969 article provides the best insights into disagreeing normatively with the incremental perspective because it discusses values.

In respect to empirical theory, we believe the starting point for assessing empirical theory is accuracy, which can be thought of in terms of being accurate in explaining and predicting budgetary phenomena. Although, as discussed above, incrementalism has been challenged in regard to the correctness of the empirical work attempting to explain budget outcomes statistically, even critics tend to confirm that budget proposals and appropriations are generally incremental in that changes are made in proposed budgets during budget processes and that appropriations change less than 30% from year to year. We find the efforts by Davis, Dempster, and Wildavsky to predict budgetary outcomes credible, and we are not aware of any published criticisms denying their general accuracy, although Padgett (1980) offers a complementary competing theory. Criticisms of the empirical work that seem most correct tend to raise questions about the degree of accuracy of incrementalism and not its basic correctness.

Incrementalism, in the authors' opinion, seems to offer reasonable perspectives in all four kinds of arguments. It is generally descriptively accurate, analytically logical, reasonable as a normative perspective, and empirically correct in general terms. In our opinion, incrementalism tends to fit observable budgeting better than any other budget theory. In addition, critics of incrementalism themselves often express a basically incremental perspective (Gist, 1982). Although incrementalism continues to be the dominant perspective in budgeting writings, it has many specific shortcomings. Over the long run, scholars who begin from that perspective are far better off acknowledging incrementalism as their starting point and seeking to improve it rather than denying its accuracy and wasting time and energy attacking it as a matter of fashion.

CONTRIBUTIONS TO FUTURE RESEARCH

Incrementalism can contribute to future research by providing a unifying perspective, a common vocabulary, and a series of problems. In our opinion, the development of budget theory, especially positive budget theory, can be best advanced by carefully considering, refining, and confronting incrementalism through looking at behavioral continuity and change in actual budgeting. Efforts along these lines would allow researchers to develop budget theory on the basis of earlier work rather than to stagger haphazardly from one fashionable budget theory to another or, even worse, to continually rediscover the same budget theory in slightly different verbal formulations previously considered abandoned. Rather than believe incrementalism to be discredited, we have discovered that its discrediting is more assumed than proven. We reached this view after reviewing the literature as objectively and as carefully as we could. In this view, we find some published support (Pitsvada & Draper, 1984; White, 1994). At the very least, incrementalism identifies variables useful in future budget research, especially those that explain and predict budgets, e.g. increments, political strategies, the preferences of individuals and groups, decision-making perspectives, and changes in political administrations. Incrementalism may prove most useful by providing a unifying perspective, a common vocabulary, and a series of problems for study.

As a unifying perspective on budgeting (as a decision-making process), incrementalism seems capable of comprehending all but two of the substantive topics of this edited work; the exceptions are the garbage can model and the interpretative approach. The other proposed topical treatments in this volume comprehend organizations, processes, and economic logic as they apply to budgeting. Incrementalism is clearly about organizations and their related political processes. The organizations include the political organization of legislatures into committees, the administrative organization of governmental operations into agencies, and the political organization of human preferences into interest groups and political parties. The political processes especially primarily comprehend those concerned with the mobilization of political support but also include ones associated with mutual partisan adjustment. Also, incrementalism relies heavily on economic logic in analyzing decision making (Schick, 1969; Wildavsky, 1974, xx). Budget theories may compete, but also in many cases, they may complement one another. The use of a unifying perspective among many budget theories can help those laboring in the vineyards of budget theory see where they and others are with their work. Also, it is interesting to note that none of the chapters in this volume argues for pursuing budget theory from the perspective of rational decision making.

With incrementalism as a unifying perspective, budget theory also has a common vocabulary. Common terms include *increment*, *calculations*, *strategies*, *process*, and *outcome*. A common vocabulary allows for, but does not guarantee, better communication among budget researchers.

Incrementalism and constructive criticisms provide a series of problems appropriate to budget theory. Explaining and predicting increments is the most obvious problem for budget theory. Others include further exploration of roles, strategies, calculations, competition, conflict, and budgetary norms. Other areas of further work based on constructive criticisms that suggest going beyond a governmental setting and into other aspects of governmental settings include interest group activities, the effects of budget outcomes, voting behavior, chief executives, legislative processes, and analysis of the outcomes of budgeting decisions. Incrementalism's long-running capacity to generate research hypotheses worth examining can be seen in a recent article that examines incremental budgeting, consensus, and dissensus (Jones, True & Baumgartner, 1997).

One ironic outcome from the movement of budget theory away from incrementalism is that while most of the theories use the same basic perspective as incrementalism they have moved away from the grounding in the study of American politics from which incrementalism emerged. Instead of looking toward the study of American politics for understanding American budgets, the sources of inspiration for later budget theories seem, on the surface, to be economics, organization theory, and sociology. Although Wildavsky drew in ideas from other areas, he argues, and we agree, that budgeting is at the heart of the political process and that changing one involves changing the other. Our enthusiasm for explanatory and predictive theories of budgeting is tempered by the apparent difficulty of developing explanatory and predictive theories of politics. Both seem to be worthwhile but also very difficult.

Other areas of great potential interest for advancing budget theory include looking earlier and later in the budget process, specifically looking more into aspects of preparation, budget implementation, and audit and review activities. Work has begun in some of these areas beyond the typical incremental focus on appropriations, e.g. budgeting of supplemental appropriations (Forrester, 1993).

Furthermore, we see incrementalism as the best starting point for advancing qualitative descriptive, analytical theory, and normative theory arguments in regard to public budgeting. The advantages of starting from incrementalism include it being well-known and reasonable.

CONCLUSION

Incrementalism is an old budget theory that is still good. It can contribute to the development of budgetary theory by providing a unifying perspective, a common vocabulary, and a series of problems to investigate in public budgeting. Despite being unfashionable, the theory has produced some of the most accurate empirical work explaining and predicting budgets. Finally, attention to incrementalism highlights the possibility that one direction to look for developing budgeting theory is politics.

REFERENCES

Bailey, J. J., & O'Connor, R. J. (1975). Operationalizing incrementalism: Measuring the muddles. *Public Administration Review*, *35*(January/February), 60–66.

Berry, W. D. (1990). The confusing case of budgetary incrementalism: Too many meanings for a single concept. *Journal of Politics*, *52*(February), 167–196.

Burkhead, J. (1956). *Government budgeting*. New York: John Wiley & Sons, Inc.

Davis, O. A., Dempster, M. A. H., & Wildavsky, A. (1966a). A theory of the budgetary process. *American Political Science Review*, *60* (September), 529–547.

Davis, O. A., Dempster, M. A. H., & Wildavsky, A. (1966b). On the process of budgeting: An empirical study of congressional appropriations. *Papers on Non-Market Decision Making 1*, pp. 63–132.

Davis, O. A., Dempster, M. A. H., & Wildavsky, A. (1971). On the process of budgeting II: An empirical study of congressional appropriations. In: R. F. Barnes, A. Charnes, W. W. Cooper, O. A. Davis, & D Gilford (Eds), *Studies in Budgeting, Vol. 11 of Studies in Mathematical and Managerial Economics* (Chap. 9, Henri Theil, Ed.). Amsterdam: North-Holland Publishing.

Davis, O. A., Dempster, M. A. H., & Wildavsky, A. (1974). Towards a predictive theory of government expenditures: U.S. domestic appropriations. *British Journal of Political Science*, *4*(October), 419–452.

Dempster, M. A. H., & Wildavsky, A. (1979). On change: Or, there is no magic size for an increment. *Political Studies*, *27*(September), 371–389.

Dempster, M. A. H., & Wildavsky, A. (1982). Modelling the U.S. federal spending process: Overview and implications. In: R. C. O. Matthews & G. B. Stafford (Eds), *The Grants Economy and Collective Consumption*. New York: St. Martin's Press.

Dempster, M. A. H., & Wildavsky, A. (1986). Toward a comparative theory of budgeting process. In: A Wildavsky (Ed.), *Budgeting: A Comparative Theory of Budgetary Processes* (2nd ed.). New Brunswick, N.J.: Transaction, Inc.

Forrester, J. P. (1993). The rebudgeting process in state government: The case of Missouri. *American Review of Public Administration*, *23*(June), 155–178.

Gist, J. R. (1974). *Mandatory expenditures and the defense sector: Theory of budgetary incrementalism*. Sage professional papers in American politics, Vol. 2. Beverly Hills, Calif.: Sage.

Gist, J. R. (1982). "Stability" and "competition" in budgetary theory. *American Political Science Review*, *76*(December), 859–872.

Hoole, F. W., Job, B. L., & Tucker, H. J. (1976). Incremental budgeting and international organizations. *American Journal of Political Science*, *20*(May), 273–300.

Jones, B. D., True, J. L., & Baumgartner, F. R. (1997). Does incrementalism stem from political consensus or from institutional gridlock? *American Journal of Political Science*, *41*(October), 1319–1339.

LeLoup, L. (1978). The myth of incrementalism. *Polity*, *10*(summer), 488–509.

Meyers, R. T. (1994). *Strategic budgeting*. Ann Arbor: University of Michigan Press.

Natchez, P. B., & Bupp, I. C. (1973). Policy and priority in the budgetary process. *American Political Science Review*, *67*(December), 951–963.

Olson, D. R. (1987). The incremental budgetary theory revisited: An empirical study of congressional appropriations. Ph.D. diss. Ann Arbor: University Microfilms International.

Padgett, J. F. (1980). Bounded rationality in budgetary research. American *Political Science Review*, *74*(June), 354–372.

Pitsvada, B. T., & Draper, F. D. (1984). Making sense of the federal budget the old fashioned way: Incrementally. *Public Administration Review*, *44*(September/October), 401–407.

Rubin, I. (1989). Aaron Wildavsky and the demise of incrementalism. *Public Administration Review*, *49*(January/February), 78–81.

Schick, A. (1966). The road to PPB: The stages of budget reform. *Public Administration Review*, *26*(December), 243–258.

Schick, A. (1969). Systems politics and systems budgeting. *Public Administration Review*, *29*(March/April), 137–150.

Schick, A. (1983). Incremental budgeting in a decremental age. *Policy Sciences*, *16*(September), 1–25.

Schick, A. (1994). From the old politics to budgeting to the new. *Public Budgeting & Finance*, *14*(spring), 135–145.

Stanford, K. A. (1992). State budget deliberations: do legislators have a strategy. *Public Administration Review*, *52*(January/February), 16–26.

Tucker, H. C. (1982). Incremental budgeting: Myth or model. *Western Political Quarterly*, *35*(September), 327–338.

Wanat, J. (1974). Bases of budgetary incrementalism. *American Political Science Review*, *68*(September), 1221–1228.

White, J. (1994). (Almost) nothing new under the sun: Why the work of budgeting remains incremental. *Public Budgeting & Finance*, *14*(spring), 113–134.

Wildavsky, A. (1961). Political implications of budgetary reform. *Public Administration Review*, *21*(autumn), 183–190.

Wildavsky, A. (1964). *The politics of the budgetary process*. Boston: Little, Brown and Company.

Wildavsky, A. (1966). The political economy of efficiency: Cost-benefit analysis, systems analysis, and program budgeting. *Public Administration Review*, *26*(December), 292–309.

Wildavsky, A. (1974). *The politics of the budgetary process* (2nd ed.). Boston: Little, Brown and Company.

Wildavsky, A. (1979). *The politics of the budgetary process* (3rd ed.). Boston: Little, Brown and Company.

Wildavsky, A. (1984). *The politics of the budgetary process* (4th ed.). Boston: Little, Brown and Company.

Wildavsky, A. (1988). *The new politics of the budgetary process*. Glenview, Ill.: Scott, Foremen and Company.

Wildavsky, A. (1992). *The new politics of the budgetary process* (2nd ed.). New York: HarperCollins Publishers.

Wildavsky, A., & Caiden, N. (1997). *The new politics of the budgetary process* (3rd ed.). New York: Longman.

Williamson, O. A. (1967). A traditional theory of the federal budgeting process. *Papers in Non-Market Decision Making 2*, pp. 71–90.

3. WINDOWS OF OPPORTUNITY: TOWARD A MULTIPLE RATIONALITIES MODEL OF BUDGETING

Kurt Thurmaier and Katherine Willoughby

ABSTRACT

This chapter explains a multiple rationalities model of public budgeting. The first section reviews the stages heuristic model of public policy-making and its applicability to a process model of budgeting. Then, Kingdon's agenda setting model is compared with Rubin's real time budgeting perspective, given their extension of aspects of the stages heuristic model to budget practice. The next section considers a multiple rationalities model of budgeting that focuses on decision-making at the micro-level of budgeting. Past experimental efforts and new evidence from ongoing research about budgeting decisions are described here. Past and present research focusing on micro-level budgeting decisions supports development of a model of multiple rationalities that explains budgeting behavior as interpretive and opportunistic. This model recognizes a more sophisticated decision activity on the part of budgeters than traditionally provided by proponents of the budget-maximizing bureaucrat (economics) or incrementalism (political science). The conclusion provokes new thought about future research in this area.

Evolving Theories of Public Budgeting, Volume 6, pages 29–54.

ISBN: 0-7623-0790-0

INTRODUCTION

Former Governor of Georgia (1983–1991), Joe Frank Harris, now a Distinguished Executive Fellow at Georgia State University in Atlanta, explained to a class of graduate students in public administration that to be successful in public budgeting you must understand the four Ps, "people, programs, process, and politics."[1] His long career in state government, 18 years in the General Assembly, his eight year tenure as Chairman of the House Appropriations Committee, and then his service for two terms as Georgia's chief executive certainly provide him with a deep reserve of wisdom related to state budget practice. His adage about the four Ps symbolizes the dynamic nature of such activity. One must understand the players involved in budgeting, the elected officials, bureaucrats, staff and clients of public services. As well, this component includes the roles and influences of the media, other interest groups, and the general public or at least, "public perception." One must understand the programs – especially, Harris noted, their correct title and organizational location – as well as their relationships to and with other programs, divisions, departments and governments. He continued by emphasizing that an understanding of process is vital to budget success; that is, the timing of requests is critical to getting your program funded. "If you are at the legislature at the beginning of the session to make a plea for your program for the first time, you are too late," he said. The politics of budgeting involves the constituent-driven actions of elected officials to support spending for services, programs, and infrastructure. Politically savvy budgeters recognize that success involves not only knowing when to strike regarding an issue, but understanding the power of certain actors in the process to secure funding for a spending item.

In one sentence Governor Harris grasps the temporal characteristics of the budget process model, from both a macro- and micro-budgeting perspective. From a macro-budgeting perspective, budgets evolve over time; passing a budget is the culmination of decisions made over time and at one point in time; a decision in and of itself is but one component in the evolution of future budgets. From a micro-budgeting perspective, the routine of a budget cycle fosters ritualistic activities among those people involved in budgeting. Accurate timing of input into these activities can be a learned or intuitive behavior on the part of budget players.

This chapter describes the process approach to public budgeting. This theoretical perspective recognizes that to be successful in public budgeting you must distinguish budgeting activities and behaviors at different stages or levels. The budget cycle itself has a beginning, middle, and end. And like a public policy, a budget is a work in progress, an evolution representative of past practice, current concerns, and anticipation of future needs. Successful budgeters understand the importance of interpreting what information comes from whom, and the appropriate time for making their case regarding future spending. The purpose of this chapter is to investigate and explain budgeting decisions at the micro-level; to understand the relationships between and among budget cycles, information streams, budgetary roles and individual choices that culminate in a final spending plan.

To explain the budget process approach, we refer to traditional consideration of policy-making in the United States as presented by various political scientists (Anderson, 1975; Jones, 1984; and Pressman & Wildavsky, 1973). Considering the nexus between policy-making and budget making is fundamental to understanding the process approach to budgeting. Consequently, we assess John Kingdon's (1995) model of the policy agenda-setting process and Irene Rubin's (1993) "real time budgeting" (RTB) for their contributions to a modern theory of public budgeting. Like most process models, these focus on macro-level budgeting. We next move to research which focuses on the micro-budgetary level, presenting research specific to individual budgetary decision-making that contributes to a multiple rationalities model of budgeting. This discussion is followed by a consideration of potential research initiatives.

THE STAGES HEURISTIC MODEL: CONTRIBUTIONS AND FAILINGS

Of traditional theories of policy-making in the United States, the stages heuristic model rather uniquely explains the policy-making *across* institutions. This model presents policy outputs as a product of activities in subsystems and not necessarily as a product of history or learned behavior within a given institution. This model departs from traditional incrementalism that focuses on bargaining among specific partisan actors. Rather, this model recognizes stages of policy-making that include agenda setting, policy formulation, implementation, and evaluation. Anderson (1975) describes policy problems as private (or individual) versus public. He explains how a policy problem surfaces and moves from the private to the public arena and then becomes a governmental problem. This model promotes a more realistic interpretation of policy-making than previously provided by institutional theory. "By shifting attention to the process stream the stages model has encouraged analysis of phenomena that transcend any given institution" (Jenkins-Smith & Sabatier, 1994, 176).

Jenkins-Smith and Sabatier (1994) discuss a number of problems with the stages heuristic model. First, they note that it does not attempt to serve as a causal model. There is little explanation of the evolution of policy from one stage to the next. Second, they note that in the real world decisions are not made in sequence as suggested by the model. Third, they claim that the model concentrates on the political arena or activities of elected officials in subsystems, while ignoring lower level decision-making activities and influences.

There is another problem with the applicability of the stages heuristic model to modern policy-making and particularly the practice of budgeting. For example, Anderson (1975, 1978) states that "probably few officials make decisions solely on ideological grounds." We consider this an artifact of past theory, given modern capabilities related to communication and technology, as well as the additional complexities of modern budgeting. Today much policy-making activity in the agenda-setting subsystem seems to be based heavily, if not predominantly, on ideological grounds (e.g. the Republicans' *Contract with America*). There is greater blurring of private and public problems in the modern states; we spend a great deal of time today on problems that in the past would easily have been categorized as individual or personal and not public or governmental. For example, under new welfare reform initiatives, some states now provide cars to workfare clients so they can get and keep a job.

The greater frequency of "spectacular" events – and the projects of such on television, by radio, and through the internet – has also led to a more difficult decision environment regarding which problems warrant governmental attention. Finally, the advance of technology, medical and otherwise, has contributed to the surfacing of problems for which there is no precedent concerning past practice. For example, recent stories about multiple births (mother gives birth to seven babies at one time in Iowa; mother-to-be expecting nine babies in Mexico) have generated heated debate in the media regarding the responsibilities of the community/government in supporting infants that the family cannot support themselves. This is an example of phenomena which just a few years ago was naturally and medically impossible and for which there is no history regarding past practices. Such changed environment is as much a function of population growth as greater and quicker access into the subsystem of agenda-setting. Ultimately, because of this modern and complex environment, research about budgeting must be much more diligent in defining the budget environment, budget actors, role orientation, and the intrinsic and extrinsic variables that influence final decisions about spending.

AN OPPORTUNISTIC MODEL OF BUDGETING

We explain a process model of budgeting by borrowing a bit from several theories of policy-making and budgeting as well as incorporating research about individual budget decision strategies. The next section begins by integrating the

RTB model with Kingdon's model of the policy process to illustrate a framework for realizing the nexus between the budget and policy processes that has evolved in governments in the United States.

Kingdon's Model of Policy-Making

Kingdon (1995) applies a combination of the process model and a garbage can approach to agenda setting in the public policy process in the United States. The garbage can model (GCM) of decision-making, unlike traditional rational choice theory, is not normative, linear, sequential, or comprehensive. According to this model, choice opportunities consist of problems and solutions that are generated by participants. What is available in a choice opportunity depends upon "the mix of [garbage] cans available, on the labels attached to the alternative cans, on what garbage is currently being produced, and on the speed with which garbage is collected and removed from the scene" (Cohen et al., 1972, 2). The structure of decision-making in this model involves the flow of relatively independent streams through the system. Decision outcomes are dependent on: (a) the coupling of these streams when a choice opportunity presents itself, and; (b) the choices available in the garbage can.

While keeping the basic logic of the model, Kingdon alters the types of streams involved and essentially converts the choice stream into two types of decision opportunities. His model has three important features: multiple decision streams, two clusters of decision actors, and two types of decision opportunities. Kingdon distinguishes between the governmental agenda and a decision agenda. The governmental agenda is structured by politics and problems. Kingdon puts a garbage can spin on the traditional policy process notion that people identify problems, generate solutions, and lobby in the political process to get them adopted. While it is true that some policy actors are active in all three areas of the policy process, he argues that there are two relatively discrete groups of policy actors (visible and hidden clusters) which tend to dominate a decision area.

Kingdon treats each of the decision areas (decisions about problems, decisions about policy alternatives, and political decisions) as separate streams in the policy process. The likelihood of significant shifts in policy choices (in contrast to incremental policy changes) increases greatly when all three streams converge and actors have a "window of opportunity" to change the current policy. As in the GCM, not all problems find solutions, and not all solutions fit problems receiving attention by other policy actors. "None of the streams are sufficient by themselves to place an item firmly on the decision agenda. If one of the three elements is missing – if a solution is not available, a problem cannot be found or is not sufficiently compelling, or support is not forthcoming from the political stream – then the subject's place on the decision agenda is fleeting. The window may be open for a short time, but if the coupling is not made quickly, the window closes" (Kingdon, 1995, 187).

Action taken during one open window can set principles that guide future decisions within a policy area, or principles that spillover into other areas. Kingdon emphasizes that most policy changes occur gradually, but landmark legislation or precedent-setting executive decisions can establish new principles that leave the policy area changed. Subsequent incremental changes that occur in the policy area then do so from a new point of origin. Even if the immediate effects are not dramatic, the importance of these events "lies in their precedent-setting nature" (Kingdon, 1995, 200).

Clusters of Actors

Extending the stages heuristic model of policy-making, Kingdon divides the policy community into visible and hidden clusters of actors. The visible cluster receives a lot of press and public attention and includes the president, high-level political appointees, and prominent members of Congress. It also includes the media, and such elections-related actors as political parties and campaign organizations. The visible cluster, not surprisingly, dominates the political stream. Visible decision makers take soundings from organized interests, and try to respond to the national mood regarding a policy area. More often than not, the national mood acts as a constraint on political decisions, limiting which problems and solutions can be addressed by the political process. Swings in the national mood create opportunities for a new set of problems or solutions to be given attention. This may be the consequence of elections that change the composition of Congress (as in 1994), put a new president and administration in charge (as in 1992), or both (as in 1980). The visible cluster of actors determines which problems move from the governmental agenda to the decision agenda. Yet they turn to the hidden cluster of actors for the set of alternatives that address the problem.

The hidden cluster generates policy alternatives and includes policy specialists such as academics and researchers, career bureaucrats, congressional staffers, and lower-level administration appointees. Career bureaucrats (and other hidden cluster actors) take their cues from the elected officials as to what items will be on the agenda for consideration. Within their "network," policy specialists try out their policy ideas on each other, measuring problems, refining and recombining solutions as necessary to make sure their alternatives to problems are ready for the decision agenda when the time is right. Much of the work of this cluster is done in the agency planning and budget shops, especially analyzing the feasibility of an idea relative to a budget constraint.

For Kingdon (1995) the heart of the policy process is "the policy primeval soup," into which are dumped the solutions to various problems under discussion in the policy community. Proposals that "surface" for serious consideration must meet criteria that include technical, political, and *budgetary* feasibility. "Thus the selection system narrows the set of conceivable proposals and selects from that large set a short list of proposals that is actually available for serious consideration" (Kingdon, 1995, 20). If elected officials are receptive to policy alternatives, then the specialists push their ideas. If not, they shelve the proposals and wait for a new administration or a new mood in the legislature (Kingdon, 1995, 44).

Kingdon (1995, 93) notes that there are two types of choice opportunities in the policy process: predictable and unpredictable "windows of opportunity." Policy changes happen during these windows of opportunity when and if the three streams come together. In that event, a problem is recognized, a solution is available, the political climate is receptive to change, and other constraints do not prohibit action. Swings in national mood brought about by crisis or turmoil are unpredictable windows of opportunity. Energy policy changes that came after the 1974 OPEC oil embargo are an example of this situation. The annual budget development process is an example of a predictable window.

Rubin's Real-Time Budgeting

Although she does not attribute her real-time budgeting model (RTB) to the idea, Rubin's (1993) model of budgetary politics has many features characteristic of both the garbage can model and Kingdon's policy process model. Rubin's RTB model begins with the notion that budgeting is not politics, but rather involves a particular set of political decision-making. It is open to the economic and political environments and must be able to cope with changing exogenous factors. The wide variety of actors with different goals, agendas, and resources only compounds the need for budgeting to be flexible and adaptable to changing circumstances. Such flexibility is provided by a decision-making structure that incorporates five distinct, loosely coupled, decision streams, each with its own cluster of decision makers and its own "politics of budgeting."

decision-making in the revenue stream is permeated with the politics of persuasion. It answers the question: who will pay how much? The principal constraint for revenue decisions is the technical estimate of the revenue base. Yet, as Majone (1989) notes, constraints can be opportunities for policy entrepreneurs. Thus, in Rubin's model, the major decisions in this cluster concern whether and how to alter the revenue base constraints with changes in taxes and tax policy.

decision-making in the expenditure stream is characterized by the politics of choice. The first constraint is the technical estimate of the base budget, which is increasingly a political decision itself. Consider modern debate over what constitutes the base budget of the federal government (dollars spent in the previous year or the "baseline" budget). The large number and variety of actors involved in this cluster are in keen competition to influence the relative allocation of revenues across competing purposes. The goal of the actors is to reorder expenditure priorities or preserve the current order of priorities, depending upon their position in the base budget.

Some of the broadest political engagements involve decision-making in the balance stream. Decisions about budget balance (how it is defined, whether to balance, and how to balance) create the politics of constraints. It is an interactive process between revenue and expenditure estimates, but fundamentally, it is linked to decisions about the scope and role of government. Debates in the 104th Congress illustrate this point well; for example, much of the debate on welfare reform was cast in terms of reducing the role of the federal government in welfare administration concomitant with capping federal expenditures to help balance the budget.

decision-making in the budget execution stream, in contrast, is viewed as much more technical in nature, and is characterized by the politics of accountability. Important questions in this cluster concern how precisely the budget plan will be followed, what deviations will be allowed, and which policy parameters cannot be violated. Changes in environmental factors often trigger implementation changes, yet budget means (programs and policies) are seldom, if ever, value neutral. Thus, depending upon the scope of changes, rebudgeting may involve considerable policy discussions. Furthermore, significant rebudgeting affects the base budget for the ensuing year, which affects both the expenditure and budget balance streams.

decision-making in the budget process stream is concerned with the politics of how to make budget decisions, and especially, who decides. Rubin argues that who participates, as individuals and groups, influences budgetary outcomes. Conflict in this stream involves the balance of decision-making power between the separate branches of government (executive versus legislature, and in modern times, the judiciary), and between the citizen taxpayers and the government officials who decide allocations. Who holds hearings, when in the process and whether they are open to statements by the general public are only a few of the issues here. This argument follows Diesing's (1962) concept of political rationality. Effective political decisions allow society to express the relative social valuations of government programs. Political rationality creates decision-making structures which balance the need for diverse viewpoints to be represented in the social debate with the need to reach some decision after the debate has occurred.

The sequential aspect of policy-making provided by the stages heuristic model dissolves in the RTB model that explains budgeting as nonlinear. Rather, budgeting requires continual decision adjustment in each stream given responses to decisions and information in other streams and changes in the political or economic environment. The streams are semi-autonomous, yet interdependent, because key information links them together. Also, a critical feature of RTB is the interruptibility of cluster decision-making, a feature required by the non-sequential timing and different decision-making intervals in the various clusters. The result is that decision blocks in one stream do not have to interrupt the rest of the clusters.

Thus, the treatment of time as a constraint is much different in the budgeting model than the policy model. In Kingdon's model of the policy process, *time* is less important than *timeliness*. In contrast, time is a real constraint in the RTB model. "Budgeting has a bottom line and a due date, which distinguishes it from many other political decisions" (Rubin, 1993, 272). The budget must be passed at some point, and the budget process as a whole is always working toward that deadline. Yet, timing is also important in the RTB model, since the various actors must be able to interrupt the work in their cluster, either to fetch necessary information from other clusters, or to revisit previous decisions in light of changing environmental conditions.

There are two aspects of Rubin's model that require more treatment to give it the full body of a systems framework. First, although Rubin notes that budgets have deadlines, the RTB does not explicitly deal with the deadlines and their impact on budgetary decision-making. Second, the RTB model, like the stages heuristic model, remains an essentially macro-level, descriptive model of budgeting. Rubin calls for others to explore the macro–micro-budgeting links further. Our exploration results in a multiple rationalities model of budgeting that is presented below.

TOWARD A MULTIPLE RATIONALITIES MODEL OF BUDGETING

Historical Perspective

Wildavsky (1992) has suggested that the complexity of modern budgeting precludes development of a comprehensive theory. Kiel and Elliott (1992, 154) agree that perhaps the best that budgeting research can uncover is "multiple rules of thumb or heuristics that capture varied processes and outcomes." Their work considers federal outlays and annual rates of change in outlays by category. They believe that budgeting is too complex, given the influences of time, change, and variation to be able to develop predictive models of budgetary outcomes. They suggest that "the development of theory necessitates an initial identification of patterns in phenomena as a means of explanation and prediction (Kiel & Elliott, 1992, 151). Howard (1973), in his classic study, *Changing State Budgeting*, adds that, "the most critical calculations in budgeting entail judgment, not mathematics." We believe there exists a growing body of research that analyzes such patterns and calculations of micro-level budgeting that helps to shape our multiple rationalities model.

Experimental Efforts to Understanding Budgeting Rationality

Results from these studies lend credence to a micro-budgeting model of multiple rationalities and opportunistic budget strategies. For example, early experimental approaches to the study of budgetary decisions include Stedry's (1960) consideration of the roles of motivation and expected reward related to budgeters' success. Results from this experiment indicate that the best performers were subjects who received explicit "high performance" expectations and then set their aspiration level. The worst performers were those who also received "high performance" expectations yet had already established their aspiration level. Stedry (1960, 90) explains that the worst performers may have set low personal standards of performance and, when confronted with high expectations for success, became discouraged and simply failed to meet such expectations. Stedry's work is noteworthy because of the unique focus on the role of expectation (information) on budget success.

Bretschneider et al. (1988) use a controlled experimental setting to study the informational influences on the decisions of budgeters. In this case, graduate students in public and business administration at Syracuse University were provided with a case description of a hypothetical, medium-sized city. Students were subjected to several "treatments" – the first involving policy role (either aligned with the council and a retrenchment perspective, or with the mayor and an expansion perspective); the second involving the accuracy of past revenue forecasts (consistently over-forecasted, under-forecasted, or oscillating). Given their role orientation and certain information, subjects then provided a maximum expenditure ceiling for the upcoming fiscal year.

Results from this experiment provide evidence of the influence of policy on the spending orientations of budgeters. While the influence of forecast bias (information) on judgment policy is less telling in terms of statistical significance, the authors find that subjects were more likely to compensate for historical over-forecasting than for under-forecasting by reducing the ceiling. From this they surmise that budgeters are able to make adjustments in their spending recommendations, "albeit imperfectly," based on historical financial data (Bretschneider et al., 1988, 318).

Similar in terms of incorporating simulation, several studies have utilized social judgment analysis to illustrate mathematically and graphically the spending policies of budgeters. Stewart and Gelberd (1976) use this method to determine the decision-making patterns of city council members, the city manager, and specific business and environmental interest group members from Boulder, Colorado. Results from this research indicate that council members and the manager were consistent in their initial judgments concerning allocation plans, and their spending policies remain consistent with their own voting records. However, these budgeters made poor predictions as to the judgments of the interest group members. The authors believe that most likely, the council members and manager simply lacked an understanding of the judgment policies of the interest group members. These results emphasize the interpretive strategies of budgeters and affects on spending decisions.

McCaffery and Baker (1990) assess the decision strategies of 34 administrators enrolled in an executive Master of Public Administration program using a variety of subjective and objective measurements involving recall and simulation. Results from this research indicate that most budgeters consider factors other than "straight workload considerations" as influencing their choice of a solution to what the authors consider purely technical problems. When comparing judgment styles, the authors find that budgeters categorized as more contextually oriented (choosing less technically correct solutions to resource allocation problems) engage a broader perspective concerning decision alternatives than their more technically oriented counterparts. Alternatively, the more technically minded budgeters (choosing the technically correct solution in the simulation) interpret their "decisional leeway" as limited. Accordingly, those exhibiting this orientation chose solutions characterized as more certain and less risky than contextually oriented budgeters. The authors find that the more technically correct the choice of solution, the greater the perception on the part of the budgeter that penalties would be imposed should workload increases not be covered sufficiently. Finally, the authors find that the more practical or contextual the decision orientation, the less the budgeter's sense of responsibility to the clientele. The significance of this research lies in its attempt to conceptualize differences between analytically and intuitively oriented budgeters in terms of choice behavior. Also, this work compares subjective and objective decision strategies.

Similar to Stewart and Gelberd, Willoughby (1993a, b) and Willoughby and Finn (1996) take advantage of multiple criteria modeling to analyze the spending decisions of state government executive (131) and legislative (89) budget analysts about hypothetical agency budget requests. Findings indicate both the weight that analysts ascribe to certain decision cues, as well as the manner in which they interpret the cues. Specifically, results show that analysts adhere to several factors when making decisions about agency spending plans. The largest group of analysts from both branches fall into the "mixed-value" decision orientation, characterized as dependent upon political and economic (not incremental) factors when making spending recommendations. Willoughby and Finn (1996, 544), having replicated Willoughby's (1993a, b) work regarding executive analysts, conclude that "in light of their heavier reliance on economic factors and broader consideration of political factors, legislative budget analysts can be characterized as more objective than executive budget analysts." These pieces illustrate the interpretive behavior of budgeters, specifically their calculation of budget cues when involved in a routine activity of their position.

Thurmaier (1995a, 449) examines the conditions in which pre-service and practicing budgeters "override an economic imperative with a political imperative for a specific budget-decision item." In this case, subjects were presented with information about the library system in a hypothetical urban city. Treatments included different fiscal forecasts, political memos from the mayor, and information presentation. Subjects analyzed information, made an initial funding decision, received new information and then made a final budget recommendation. Thurmaier, similar to Willoughby (1993a), finds that analysts *interpret* the political environment when making spending decisions. While political factors often override the economic ones in a decision, he (1995a, 459) notes that, "not all of the political cues carry a political imperative; those that do not are subordinated to economic imperatives." Also similar to Willoughby (1993a), Thurmaier finds that novice budgeters seem less sure of their decisions, that is, they are more likely to change their decision with new information. It is interesting that he also finds that information order was important in shaping not only initial recommendations, but also the probability that budgeters would change their recommendations with new information. In short, analysts were likely to remain anchored (Khaneman, Slovic & Tversky, 1987) to their initial recommendation unless faced with a subsequent political imperative to change.

These works are important for they illustrate the interpretive decision behavior of budgeters of various role orientations. From this body of work, we conclude that budget decisions involve multiple factors and budgeters tend to apply multiple rationalities to filter various types of information received at different points in time.

Current Research Supporting a Multiple Rationalities Model

Our current research focuses on the decision behavior of analysts in central budget bureaus (CBBs). (See Appendix for research focus, sample, and protocol.) An emphasis on budget process highlights the gatekeeping role of these budgeters in the executive budget process commonly found at all levels of government in the United States. Miller (1991, 78) emphasizes that the role of the finance officer is to "help focus attention on truly relevant issues at just the right time to gain the power to interpret events important to the organization." CBB analysts have considerable influence on budgetary outcomes, serving the chief executive as powerful gatekeepers who stem the flow of budgetary requests to the chief executive and legislators and who increasingly work beyond mere financial analysis to serve the role of policy analysts (Berman, 1979; Davis & Ripley, 1969; Johnson, 1984, 1988, 1989; LeLoup & Mooreland, 1978; Gosling, 1987, 1985; Willoughby, 1993a; Thurmaier, 1992).

Historically, legislatures have delegated the vast majority of the expenditure decisions to the governor (Kiewiet, 1991; Lauth, 1992, 1077), who then delegates significant decision-making power to the budget staff in the CBB (Gosling, 1985, 1987). Gosling's (1991) study of gubernatorial policy agendas finds that governors have increased their use of budgets as policy vehicles, especially in the latter half of the 1980s. Lynch (1995) points out that analysts in state CBBs are uniquely qualified to pursue policy development and innovation because their allegiance to the governor exposes them to a global perspective of state government, yet their contact with line managers focuses their attention on specific program initiatives. Forsythe (1991, 171) observes that these budgeters "usually have a good idea of which agencies are doing well, which are doing poorly, which agencies are using resources intelligently and efficiently, and which are not."

A CBB with a policy orientation gives budgeters key roles as actors in both the budget and policy process. Although modern governors' macro-budgeting decisions are increasingly bounded by non-discretionary spending requirements (for example, for social welfare programs, formula-based education programs, Medicaid, earmarked categories like transportation, and mandated spending such as for regional hospitals and corrections), it is through the budget office that non-discretionary and discretionary policy decisions are transmitted to agencies. The budget represents an alignment of micro-budgeting decisions of career bureaucrats with the macro-budgeting decisions of elected officials.

Figure 1 illustrates this macro/micro relationship in budgeting. As noted in this figure, both macro- and micro-budgeting decisions are framed by environmental, organizational, and technical factors. Environmental factors like federal

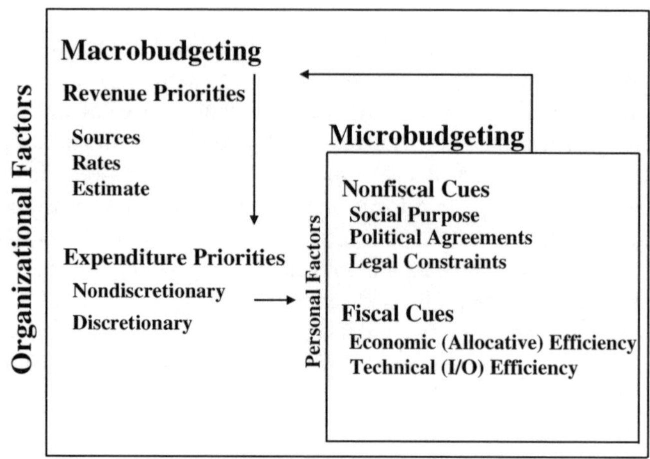

Fig. 1. The Nexus between Macro- and Micro-Budgeting Decisions: The Decision Window of the Central Budget Bureau Budgeter.

tax reform, unfunded mandates, court orders, and natural disasters can dramatically change revenue and expenditure priorities. In turn, such events can significantly affect micro-budgeting decisions – overshadowing both the nonfiscal and fiscal decision cues considered by budgeters. Organizational and technical factors relate to the capacity of decision makers to gather information, formulate decision strategies, and make informed choices. Also shown in Fig. 1, the personal backgrounds, experiences and activities of budgeters influence their understanding of the budget process, relationships with other players, consideration of budgetary cues, and prioritization of government activities.

The Budget as the Biggest, Most Predictable, Window in Town

The prominent and pivotal role of the CBB in budgetary communications flows illustrated above squares with the general characteristics of Kingdon's policy model. He (1995, 105) notes that the "budget constitutes a particular kind of problem . . . [because] the budget is a central part of governmental activity." While budget considerations can promote issues higher on the governmental agenda, more often the budget acts as a constraint because the item exceeds costs that decision makers are willing to contemplate. In either case, budgetary considerations and policy process are intertwined. Kingdon (1995, 108) points out that some of the budget constraint is perceptual because it is "subject to interpretation . . . The budget constraint can be cited as an argument against a proposal that one does not favor on other grounds, and can be side-stepped for proposals that one does favor, by underestimating their cost or ignoring their long-range cost altogether." This point is supported by the significance of the order factor in Thurmaier's (1995a) experiment.

While he says little about the specific role of the budget office in the policy process, Kingdon (1995, 26–27) places OMB in the position of being hidden presidential staff, largely devoted to alternative specification, but retaining its central gatekeeper function. "The Office of Management and Budget . . . has some enduring orientations that persist, regardless of the turnover of personnel within OMB or the comings and goings of administrations . . . Everybody in government . . . can count on OMB to be interested in cutting budgets, and, in the case of new initiatives, opting for the least expensive program possible" (Kingdon, 1995, 162). Several analyses of OMB urge a preservation of the gatekeeper function and "neutral competence" (Berman, 1979; Tomkin, 1998). The credibility of OMB has been seriously questioned when its policy role has overwhelmed the control function as occurred during the Kennedy, Nixon, and Reagan tenures (Tomkin, 1998). Kingdon (1995, 186) adds that agency budget bureaus are also part of the most predictable windows of opportunity in which policy actors have the opportunity to make significant changes to public policies and programs.

We are developing a micro model that parallels the macro model. Expanding Kingdon's notion of a decision agenda, we propose a parallel adaptation to budgeting: the budgetary decision agenda. The budgetary decision agenda is the set of issues that policymakers are actively considering for inclusion in the next budget. As with the policy decision agenda, the budgetary decision agenda is set by the visible cluster of actors, particularly the chief executive. And also following Kingdon's model, the specification of budget and policy alternatives is largely in the hands of the hidden cluster of actors. In the case of the budget window of opportunity, the key actors responsible for the coupling of problems and solutions are the CBB analysts.

In our research of state level analysts, we consider that these budgeters have a policy role when their activity involves them in discussions and decisions regarding state policies in a substantive policy area *as such issues appear on the budgetary decision agenda*. It is not necessary, however, for all budgeters in a CBB to have a significant policy role for the office to have a policy orientation. An analyst assigned to the budget of the cosmetology board, for example, would be expected to have little policy content compared to the one assigned to the Medicaid budget. Other factors that may affect the level of policy influence a single budgeter has on state budgeting include the position of the policy issue on the decision agenda, the size of the budget request, and the budget office experience of the analyst.

Our interviews with state government analysts suggest that as gatekeepers, their policy activities are concentrated in the budgetary windows of opportunity. They analyze policy alternatives that have cooked and survived in the "policy primeval soup." The analysts' input provides "flavor" to this soup. For example, one analyst explains the budgeting process as a "mix" in which analysts' recommendations are a vital ingredient that contributes to final budget and policy decisions.

> There is the revenue availability, there is a billion dollars worth of requests, there is the governor's agenda. You throw your recommendations in the pot. Then, the governor says, "I want this one, this one, and this one." An analyst's recommendations can get lost in the mix. Obviously, how you present a request, how you write it up, and the documentation that accompanies it, carries some weight regarding what the governor decides.

Bozeman and Massey (1982) emphasize that policy evaluation is useless if the political environment is not properly accounted for on the part of analysts. CBB analysts are cognizant of this fact while sifting through numerous forms of information from different sources across time – the weight that analysts give to the fiscal and nonfiscal cues illustrated in the decision window of Fig. 1 can change throughout the course of a budget process (fiscal year). Analysts verbally illustrate this interpretive notion of budgeting, indicating strategies that are dependent upon streams colliding at any given time. For example, an analyst, when asked about supporting an agency request before the governor, discloses the environmental scan of political and revenue streams common of those involved in budget preparation,

Interviewer: What distinguishes whether you fight [for an agency] or move on?

Analyst: That sometimes changes during the year, depending upon the revenue forecast. If the revenue looks bad, I may give up on it.

Interviewer: Even if you know that the department really needs it?

Analyst: If I think there is a good chance I can convince my director, and those in the governor's office, that this is a mandatory item and has to be funded, I will go to the wall for it. If there is any doubt about that, and the fiscal situation looks bad, I am not going to spend a lot of time on it.

When the budget process is structured to give CBBs a policy orientation, analysts act on behalf of the chief executive in the policy-making process.

Analysts take their cues from political actors regarding which budget decision items are on the budgetary decision agenda and which are not (Thurmaier, 1995a, b; Willoughby, 1993a, b; Willoughby & Thurmaier, 1995). When items are not on the agenda, analysts are informally delegated decision authority by the legislature and chief executive, and traditionally, their instinctive response is to "just say no."[2] When analysts receive cues that a policy issue is on the budgetary decision agenda, they are critical gatekeepers who help determine the set of legitimate alternatives. In any event, they must persuade elected officials that their recommended policy change is feasible and in accord with the chief executive's policy focus. To this end, we find among the analysts interviewed an interpretive attention to agency budget requests that almost represents agency advocacy, many expressing the sentiment that they help agencies "know what to look for." An analyst says,

> We perform for agencies in the sense that we are closer to the governor's priorities, and we know what the button bar is in terms of preferred language and style. There are times when agency personnel write something which you know the governor would like if they just wrote it in a different way. But, they do not know which terms to use and they may not emphasize the right things. Agencies are looking at things parochially. They do not know their audience as well as we do. So, we can help.

A pamphlet on the role of the state budget analyst in this state specifies that analysts, "verify need, verify cost, seek alternatives, challenge the base, and serve as an advocate for the agency when appropriate (when there is pressure to cut the base)." Similarly, one analyst admits that, "it is amazing what you can add to agency budgets and programs just by going and visiting and talking to them." Still another describes budget strategy as "finding an angle and trying to make the governor see it." Nonetheless, these budgeters understand that agency advocacy only goes so far, given the political and budgetary feasibility necessary to provide for a successful coupling of problem with solution. Ultimately, according to one analyst "you must be able to look at something and know the politics, the governor's agenda, and is it fiscally sound."

As key managers of the budgetary decision agenda, central budgeters take cues from political leaders as to what problems will be on the agenda, and then sift and hone the alternatives from the policy community to determine which alternative or alternatives will be promoted for consideration by elected officials. After all, as many analysts in our study emphasized, "the overriding priority is the governor's agenda." For this reason, other political actors in the decision-making structure are keen to get agreement from the central budget bureau, thereby surmounting a major obstacle in the policy-making process (Duncombe & Kinney, 1987; Mosher, 1952; Wildavsky, 1988; Davis & Ripley, 1969; Tomkin, 1998).

Thus, analysts are targets of persuasion (Cowart & Brofoss, 1978; Tomkin, 1998). They are subjected to various and conflicting alternatives promoted for the same problem. However, the budget director usually wants a recommended alternative and the rationale to show that the recommended alternative satisfies the three feasibility criteria.

Kingdon (1995, 139) observes that policy specialists discard many ideas because they "cannot conceive of any plausible circumstances under which they could be approved by elected politicians and their appointees. Some ideas are kept alive in the hope that the larger political climate will change, even though the ideas might not be currently in favor." Analysts, on the other hand, are criticized only infrequently for making recommendations that are responsive to policy direction, yet unpopular with the chief executive. The sensitive judgment that Veillette (1981, 67) notes is required of budgeters incorporates the budgetary politics of the issue, even while maintaining standards of neutral competence; that is, "one must have the discernment to avoid stretching a good principle to its breaking point."

This portrait of budgeting in CBBs requires further texturing. The time constraint that RTB imposes on a GCM framework redefines the portrait of organized anarchy in an important way. Unlike its counterpart in policy-making, the budget process forces a confluence of the RTB decision-making streams when the deadlines arrive to pass the budget. As the deadlines near, the CBB yokes the streams closer together. These streams need not conjoin simultaneously. But eventually, they must be integrated fully for a new budget to be approved – if only for two short moments each year (the deadline for the executive to submit a budget proposal to the legislature, and the vote by the legislature to pass the budget). In the same moments that the budget is approved, so are the policy decisions embedded in the budget.

While the analyst's role seems fairly clear during budget development, their role during the legislative session does not. One deputy director emphasized however, that analysts may play just as decisive a role during this period of budgeting. He points out that CBB analysts must be comfortable with seizing the opportunity to become involved in policy and budgeting decisions.

> If we (CBB analysts) have an opportunity to be heard in the legislative process, we stand up in a diplomatic way and say, "I disagree with you," and say why. Analysts in this office have an opportunity to do that. If an analyst does not express his or her interpretation of information when the opportunity arises, sometimes they lose that moment to stop something from happening.

Growing evidence suggests that state and local legislative bodies accept 90 percent or more of the executive's budget recommendations (Gosling, 1987; Willoughby & Thurmaier, 1995; see also Crecine 1969; Rickards, 1984; Wildavsky, 1975). Thus, the bulk of the problem-solution-politics coupling on the budget decision agenda is achieved when the executive approves the budget to be proposed to the legislature. Any of the "decided" issues can be revisited during legislative deliberations, and many are. The decision streams in the RTB model immediately disjoin once the executive's policy and budget decisions are announced. Revenue, expenditure and balance decisions will be reviewed, modified, and remodified, especially as revenue and expenditure forecasts are updated. The structure of the legislative budget process will delimit the amount of the conflict surrounding the proposals. Ultimately, the streams must conjoin again for the budget to become law, and most of the chief executive's recommendations will be affirmed.[3]

The CBB as the Vortex

CBB budgeters are well placed to monitor the various decision streams in the policy and budget process. Many analysts we interviewed support the characterization of the CBB as, "where the action is," and offering "a global perspective" or "the big picture." According to one analyst in our study, "if it is something that the agency has to do that affects policy, programs or dollars, then we will look at it." And, the powerful role of the CBB analyst has an additional benefit – less public accountability than more visible budget players. According to one, "we have exposure without having to put our necks on the chopping block all the time." And analysts are fluid participants in the process, discussing revenues with one program, expenditures with others, balance issues and execution issues with still others. One analyst describes the decision process as "very fluid so that decisions can be made and unmade at several different points."

At some point during the budget year analysts will participate in choice opportunities in each of these streams with each of their assigned agencies. They know other participants and monitor problem development and evolution specific to each.

Their links with agency policy actors (including program directors) keep them abreast of solutions in the policy primeval soup. They may participate in evaluations of solutions for other feasibility criteria but their primary role concerns budgetary feasibility. With one eye on the policy process and one eye on the budget process, they evaluate how various solutions fit with the prevailing flow of decisions and preferences of the chief executive.

The Nexus of Macro and Micro Decisions

The notion of multiple decision streams of various types is very compatible with our evidence that budgeters use a multiple rationalities approach to micro-budgeting decisions. As presented in Fig. 1, past experimental research and current field studies of CBBs and analysts suggests that budgeters' decisions are actually composites of multiple decisions of different types at various points in a budget cycle. Results suggest that analysts combine fiscal and non-fiscal decisions into rational recommendations that they can defend before the budget director and the governor. Most analysts appear to base their recommendations on evaluation of the sociopolitical context as well as the economic and technical factors affecting the agency budget problem under his or her review.

The combination of fiscal and non-fiscal decisions at the micro-budgeting level is a direct consequence of analysts' critical role at the nexus of macro- and micro-budgeting decisions. The task of harnessing agency budget developments to support macro-budgeting decisions of the governor (and the legislature) adds a second layer of complexity to the micro-budgeting decisions of these actors. They cannot make defendable micro-budgeting decisions without a substantial understanding of the key environmental variables and the disposition of macro-budgeting decisions. An analyst explains the interpretive quality of this actor's duties:

> Analysts have to have the ability to analyze, the ability to pull apart an issue, see its components and see what parts of an issue drive the other parts, what the needs are, and then be able to derive logical conclusions from that and develop logical solutions to whatever the problem is. Finally, the analyst must be able to articulate solutions, both in written and verbal form.

At the same time that budgeters in CBBs serve a vital role as the nexus between top-down and bottom-up budgeting, they are also serving as the nexus between the policy and budget processes. Much of the sociopolitical information that provides budgeters with the context for their more technical analyses relates to their consideration of the policy issues surrounding the budget agenda item. The issues Rubin (1993) enumerates in the budget balance stream (including the role and scope of government) are examples of the swirling mix of policy and budget decisions that converges on the CBB and its budgeters.

CONCLUSION

The notion of fluid participation across decisions about revenues, expenditures, budget balances, budget process, and budget execution is appealing because we have learned that *who* is budgeting *when* matters for budgetary outcomes. Problematic preferences are intrinsic characteristics of public budgeting because problems are ambiguous and capricious, solutions have chameleon qualities, and the politics of an issue depends on who is politicking.

The stages heuristic model of policy-making first provided an overview of decision-making in subsystems although with a macro-level focus. Kingdon extends this model by recognizing the role of the invisible cluster of actors in the policy process. The garbage can model provides a good picture of macro-budgeting, and also creates the context for micro-budgeting decisions. If we look into the central budget bureau situated at the nexus of macro- and micro-budgeting, and at the nexus of budgeting and policy-making, the picture is more structured than chaotic. We are left with an image of the CBB as a vortex through which flow both policy and budget decision streams. Thus, their influence is felt throughout both processes.

The advancement of sophisticated techniques to measure human cognitive activity bodes well for future study in the field. To extend this line of research, we suggest that additional and repeated applications of the laboratory approach, simulation techniques as well as face-to-face interviews be conducted to determine the spending policies of practicing public budgeters. Analysis of expert judgment is facilitated because the routine and repetitive nature of the budget cycle fosters expertise among budget players. Analyses must concentrate on decision strategies and patterns of budgeters of similar roles across different levels of government (budget analysts in central budget bureaus at the federal, state and local levels). As well, analyses must consider budgeters of different roles within one government (for example, comparison of the decisions of budget analyst behavior to those of the chief executive, agency line and staff, legislative branch members and their fiscal staff).

To complement these approaches, attempts should be made to compare budgeters' decision-making strategies or judgment orientations with the results of such activity. This may necessitate the examination of real funding recommendations for specific agencies in conjunction with judgment analysis to determine the congruence between budgeters' recorded strategies and actual decision outcomes over time. It is clear that any consideration of modern budgeting must include an assessment of the activities of the invisible cluster of actors, that is, consideration of micro- as well as macro-level decision processes. Bowling et al. (1996, 24) concur: "by returning to the micro-level motivational and behavioral analyses, we also return to comprehensive and cumulative process-oriented approaches of both Niskanen and Wildavsky that continuously influences budgetary theory-building" (for example, comparison of the decision behavior of budget analysts to that of the chief executive, agency line and staff, legislative branch members and their fiscal staff).

NOTES

1. March 5, 1997, Georgia State University, Atlanta, Georgia.
2. Sharkansky (1969, 38–49) labels this phenomenon as "contained specialization." See also, Gosling (1985).

3. The phenomenon is strongest for state and local governments. It is weakest at the national level, where sharp partisan differences between the legislative and executive branches can significantly reduce the president's influence on the budget. Aside from these alignments, the impact of the president's recommendation on the final budget passed by Congress is substantial.

REFERENCES

Anderson, J. E. (1975). *Public Policy-Making*. New York, NY: Praeger Publishers, Inc.

Berman, L. (1979). *The Office of Management and Budget and the Presidency, 1921–1979*. Princeton, NJ: Princeton University Press.

Bowling, C. J., Burke, B., Jensen, J. M., & Wright, D. S. (1996). Beyond the maximizing Bureaucrat: Explaining Variations in the Budget Preferences of American State Administrators. Paper presented at the annual meeting of the Southern Political Science Association, Atlanta, Georgia, November 7–9.

Bozeman, B., & Massey, J. (1982). Investing in Policy Evaluation: Some Guidelines for Skeptical Public Managers. *Public Administration Review*, *42*(May/June), 264–270.

Bretschneider, S., Straussman, J. J., & Mullins, D. (1988). Do Revenue Forecasts Influence Budget Setting? A Small Group Experiment. *Policy Sciences*, *21*, 305–325.

Clynch, E. J., & Lauth, T. P. (1991). *Governors, Legislatures, and Budgets*. New York: Greenwood Press.

Cohen, M. D., March, J. G., & Olsen, J. P. (1972). A Garbage Can Model of Organizational Choice. *Administrative Science Quarterly*, *17*(3).

Cowart, A. T., & Brofoss, K. E. (1978). *Decisions, Politics, and Change: A Study of Norwegian Urban Budgeting*. Olso: Universitetsforlaget.

Crecine, J. P. (1969). *Governmental Problem-Solving: A Computer Simulation of Municipal Budgeting*. Chicago: Rand McNally & Co.

Davis, J. W., & Ripley, R. B. (1969). The Bureau of the Budget and Executive Branch Agencies: Notes on Their Interaction. In: J. W. Davis, Jr. (Ed.), *Politics, Programs, and Budgets: A Reader in Government Budgeting* (pp. 66–67). Englewood Cliffs, NJ: Prentice-Hall.

Diesing, P. (1962). *Reason in Society: Five Types of Decisions and Their Social Conditions*. Urbana: University of Illinois Press.

Duncombe, S., & Kinney, R. (1987). Agency Budget Success: How It is Defined by Budget Officials in Five Western States. *Public Budgeting and Finance*, *7*(Spring), 24–37.

Forsythe, D. W. (1991). The Role of Budget Offices in the Productivity Agenda. *Public Productivity and Management Review*, *15*, 169–174.

Golembiewski, R. T., & Rabin, J. (Eds), (1983). *Public Budgeting and Finance: Behavioral, Theoretical, and Technical Perspectives* (3rd ed.). New York: Marcel Dekker, Inc.

Gosling, J. J. (1991). Patterns of Stability and Change in Gubernatorial Policy Agendas. *State and Local Government Review*, *23*(Winter), 3–12.

Gosling, J. J. (1987). The State Budget Office and policy-making. *Public Budgeting and Finance*, *7*(Spring), 51–65.

Gosling, J. J. (1985). Patterns of Influence and Choice in the Wisconsin Budgetary Process. *Legislative Studies Quarterly*, *10*(November), 457–482.

Howard, S. K. (1973). *Changing State Budgeting*. Lexington KY: Council of State Governments.

Jenkins-Smith, H. C., & Sabatier, P. A. (1994). Evaluating the Advocacy Coalition Framework. *Journal of Public Policy*, *14*(2), 175–203.

Johnson, B. (1989). The OMB Budget Examiner and the Congressional Budget Process. *Public Budgeting and Finance*, *9*(Spring), 5–14.

Johnson, B. (1988). OMB and the Budget Examiner: Changes in the Reagan Era. *Public Budgeting and Finance*, *8*(Winter), 3–21.

Johnson, B. (1984). From Analyst to Negotiator: the OMB's New Role. *Journal of Policy Analysis and Management*, *3*(Summer), 501–515.

Jones, C. O. (1984). *An Introduction to the Study of Public Policy* (3rd ed.). Belmont, CA: Wadsworth Publishing Company.

Kahneman, D., Slovic, P., & Tversky, A. (Eds) (1987). *Judgment Under Uncertainty: Heuristics and Biases*. NY: Cambridge University Press,

Kiewiet, R. D. (1991). Bureaucrats and Budgetary Outcomes: Quantitative Analysis. In: A. Blais & S. Dion (Eds), *The Budget-Maximizing Bureaucrat: Appraisals and Evidence*. (pp. 143–174) Pittsburgh: University of Pittsburgh Press.

Kiel, L. D., & Elliot, E. (1992). Budgets as Dynamic Systems: Change, Variation, Time and Budgetary Heuristics. *Journal of Public Administration Research and Theory*, *2*(2), 139–156.

Kingdon, J. W. (1995). *Agendas, Alternatives, and Public Policies* (2nd ed.). NY: Harper Collins.

Lauth, T. P. (1992). State Budgeting: Current Conditions and Future Trends. *International Journal of Public Administration*, *15*, 1067–1096.

LeLoup, L. T., & Moreland, W. B. (1978). Agency Strategies and Executive Review: The Hidden Politics of Budgeting. *Public Administration Review*, *38*(May/June), 232–239.

Lynch, T. D. (1995). *Public Budgeting in America* (pp. 166–212). Englewood Cliffs, NJ: Prentice-Hall, Inc., Ch. 6, Analytical Processes.

Majone, G. (1989). *Evidence, Argument and Persuasion in the Policy Process*. New Haven, CT: Yale University Press.

McCaffery, J., & Baker, K. G. (1990). Optimizing Choice in Resource Decisions: Staying Within the Boundary of the Comprehensive-Rational Method. *Public Administration Quarterly*, *14*(Summer), 142–172.

Miller, G. J. (1991). *Government Financial Management Theory*. New York: Marcel Dekker, Inc.

Mosher, F. C. (1952). Executive Budget, Empire State Style. *Public Administration Review*, *12*(Spring), 73–84.

Pressman, J. L., & Wildavsky, A. (1973). *Implementation*. Berkeley, CA: University of California Press.

Rickards, R. C. (1984). How the Spending Patterns of Cities Change: Budgetary Incrementalism Reexamined. *Journal of Policy Analysis and Management*, *4*(1), 56–74.

Rubin, I. S. (1993). The Politics of Public Budgeting (2nd ed.). Chatham, NJ: Chatham House.

Schick, A. (1988). Micro-budgetary Adaptions to Fiscal Stress in Industrialized Democracies. *Public Administration Review*, *48*(January/ February), 523–533.

Sharkansky, I. (1969). *The Politics of Taxing and Spending*. NY: Bobbs-Merrill.

Stedry, A. D. (1960). *Budget Control and Cost Behavior*. Englewood Cliffs, NJ: Prentice Hall, Inc.

Stewart, T. R., & Gelberd, L. (1976). Analysis of Judgment Policy: A New Approach for Citizen Participation in Planning. *American Institute of Planners Journal*, (January), 33–41.

Thurmaier, K. (1995a). Responsive and Responsible Budgeteers: An Experiment in Budgetary decision-making. *Public Administration Review*, *55*(5), 448–460.

Thurmaier, K. (1995b). The Multiple Facets of Budget Analysis: Focusing on Budget Execution in Local Governments. *State and Local Government Review*, *27*(2), 102–117.

Thurmaier, K. (1992). Budgetary decision-making in Central Budget Bureaus: An Experiment. *Journal of Public Administration Research and Theory*, *2*(4), 463–487.

Tomkin, S. L. (1998). *Inside OMB: Politics and Process in the President's Budget Office*. Armonk, New York: M.E. Sharpe.

Veillette, P. T. (1981). Reflections on State Budgeting. *Public Budgeting and Finance*, *1*(3), 62–68.

Wildavsky, A. (1988). *The New Politics of the Budgetary Process*. Boston: Scott, Foresman/Little Brown College Division.

Wildavsky, A. (1992). *The New Politics of the Budgetary Process* (2nd ed.). Harper Collins Publishers, Inc.

Wildavsky, A. (1975). *Budgeting: A Comparative Theory of Budgeting Processes* Boston: Little Brown and Company.

Willoughby, K. G. (1993a). decision-making Orientations Of State Government Budget Analysts: Rationalists or Incrementalists? P*ublic Budgeting and Financial Management*, *5*(1), 67–114.

Willoughby, K. G. (1993b). Patterns of Behavior: Factors Influencing the Spending Judgments of Public Budgeters. In: T. D. Lynch & L. L. Martin (Eds), *The Handbook of Comparative Public Budgeting and Financial Management* (pp. 103–132) NY: Marcel Dekker, .

Willoughby, K., & Finn, M. A. (1996). Decision Strategies of the Legislative Budget Analyst: Economist or Politician? *Journal of Public Administration Research and Theory*, *6*(October), 523–546.

Willoughby, K., & Thurmaier, K. (1995). The Changing Roles of Central Budget Bureaus and Analysts: A Comparison of Midwest and Southeast States. Paper presented at the annual meeting of the Midwest Political Science Association, Chicago, Illinois, April.

APPENDIX

Research Methodology

Researchers conducted extensive face-to-face interviews with analysts in each of the 11 states in the sample (Alabama, Georgia, Illinois, Iowa, Kansas, Minnesota, Missouri, North Carolina, South Carolina, Wisconsin, and Virginia). A total of 182 CBB Budgeters were interviewed during the spring, summer, and fall of 1994, with the exception of Kansas analysts who were interviewed during the summer of 1993. (Our sample includes analysts, section managers, deputy directors, directors and one Department Secretary.) Typically, a researcher visited a budget office over a period of days, conducting interviews lasting from 30 to 90 minutes. On average, interviews lasted about 45 minutes. Each session began with an account by the analyst of the length of time they had been employed in the budget office, and their previous work experiences and educational background. Interview questions included perceptions of their role in the budget process, the general orientation of the budget office factors they consider most important when reviewing agency programs and services to determine spending plans for the upcoming fiscal year, who they work for, and the most important characteristics of analysts to be effective. The core of the interview concentrated on their strategy(ies) for collecting information about agencies under their purview, their perception of agency and gubernatorial agendas, and their role in relaying information to both entities. Except in the rare instance that an analyst declined permission to be taped, all interviews were recorded and transcribed for analysis.

The sample of state governments in this study is representative of the broad range of key state characteristics in the U.S., including population, financial well-being, and political makeup. For example, Illinois is the sixth most populous state, with the fourth largest gross state product, and a citizenry with the 10th highest per capita disposable income (United States Bureau of Census, 1993). North Carolina, Georgia and Virginia are ranked 10th through 12th in population, respectively. In contrast, Kansas is the least populous state in the sample, ranked 32nd in the US, with a gross state product ranked 31st in the U.S. (and lowest in the sample). The sample includes some of the poorest states in the U.S.: South Carolina ranks 42nd, Alabama 40th and Iowa 38th.

At the time of the study, six of the 11 states had Republican governors. Three of these governors faced predominantly Democratic legislatures (in Minnesota, South Carolina and Virginia). The other three worked with split-party legislatures (Iowa, Illinois and Wisconsin). Fives states had Democratic governors and split-partisan legislatures. All but the North Carolina governor have some degree of budgetary veto power.

These states also differ in the budgetary power and organizational arrangements afforded to each governor and reflect the different and changing executive-legislative arrangements discussed by Clynch and Lauth (1991). For example, governors in six states have the primary authority for setting the revenue estimate, the powerful budget tool that frames the decisions about future state spending. The budget offices in three states (Georgia, Illinois and North Carolina) are located within the Office of the Governor. Yet the North Carolina governor is weakened because there is no executive veto power, and an arm of the legislature prepares the revenue estimate. The revenue estimate is prepared by consensus in Alabama, Iowa, and Kansas; it is a joint legislative-executive exercise in Wisconsin. In general, we believe the diversity in the sample of states allows us to draw valid and reliable conclusions about the changing roles of budgeters in state CBBs. We also welcome efforts to extend this research to other states and regions to test the external validity of the study.

4. ORGANIZATIONAL PROCESS MODELS OF BUDGETING

Mark T. Green and Fred Thompson

> If the field does not invest in behavioral theories now, the odds are good that we will suffer from collective amnesia, forgetting the relevant techniques. And having forgotten how to use a screwdriver, modelers will continue hammering (Jonathan Bendor, 1990).

INTRODUCTION

Mention of the organizational process model[1] often draws blank stares from students of public budgeting. Sometimes explanation does not materially improve comprehension. Bibliographical searches yield only a handful of references or citations. And most contemporary texts entirely overlook the topic. It is only when one cites the originators of the organizational process model that the conversation is transformed from a monologue to a dialogue.

Nevertheless, when social scientists hear Herbert Simon, Charles Lindblom, or Aaron Wildavsky's name, they tend to react in terms of contributions to a specific discipline. Many of our colleagues seem to have forgotten that these scholars and their students once constructed a powerful predictor of budgeting behavior – the organizational process model. It is the mission of this paper to remind scholars of this impressive body of scholarship.

There are several reasons for the obscurity of the organizational process model. Perhaps the most important is the widespread acceptance of the standard economic model which holds that spending outcomes are determined by the supply of, and the demand for, publicly provided goods and services, and competition – either for public office or though a Tiebout-type mechanism –

Evolving Theories of Public Budgeting, Volume 6, pages 55–81.

ISBN: 0-7623-0790-0

constrains elected officials to serve the tastes and preferences of the median voter (Thompson, 1997). Moreover, positive political theorists, especially economists but political scientists as well, have been understandably reluctant to embrace organization process models. The rhetoric employed by organizational process theorists is unfamiliar; their models are inelegant and hard to validate; their methodology appears cumbersome; and they were/are hostile to attempts to incorporate their findings into the standard partial equilibrium framework of positive political theory.

Positive political theorists are not alone in neglecting organizational process models. Institutionalist and behaviorally oriented budget scholars are equally disdainful of them. This fact is somewhat harder to explain. Perhaps the organizational process approach to expenditure analysis was a good idea with bad timing. Or, perhaps, its advocates were not aggressive enough in converting researchers to their methodology. While these are likely contributing factors, we believe that the most significant cause for the current oblivion of the model is its inherent multidisciplinarity.

Scholars preach the virtues of multidisciplinary perspectives and research, but do not often practice what they preach. Academics are rewarded for contributions to disciplinary knowledge, usually with a narrow focus on a single subfield or methodology. It doesn't make much sense to expect x while rewarding y. As a consequence, organizational process research is the victim of cross-field diffusion. Its findings and methods have been scattered to the four winds – across a variety of fields and disciplines.

What is the organizational process model? While we could provide a definition, it would be a hollow effort at this time and lack any meaning. The best way to define the organizational process model is to follow its interesting etiology across time and academic boundaries. This process of development can be divided into four distinct phases: pronouncement, formalization, differentiation, and diffusion. This article looks in turn at each of these phases in the development of the organizational process model and concludes that the time has come for scholars to revisit the model, exploring its potential for contributing not only to formal budget theory but also to other fields of public policy and management as well.

PRONOUNCEMENT OF THE ORGANIZATIONAL PROCESS MODEL

The origins of the organizational process model of budgeting can be traced to the work of three prominent scholars – Herbert Simon, Charles Lindblom and

Aaron Wildavsky. Each of these scholars worked, for the most part, separately and within an established disciplinary framework – public administration, economics, and political science – helping to craft the core of the theory. Moreover, each implicitly adopted the same unit of analysis, the decision, and explicitly relied on the power of cross-disciplinary research in shaping their unique contributions to our understanding of the budgetary process. We begin with the work of Nobel laureate Herbert Simon.

The first cornerstone of the organizational process model is Simon's now well-known theory of "bounded rationality," which was best articulated in "A Behavioral Model of Rational Choice" (1955) and later in *Models of Man* (1957) as an alternative to strong forms of the rational choice model of decision-making. Bounded rationality reflected Simon's criticism of the assumption that decision makers are utility maximizers. According to Simon decision makers cannot be perfectly rational because it is impossible to be perfectly rational. One cannot hope to calculate the probabilities of every relevant outcome under every alternative included in the set of mutually exclusive and severally exhaustive options, let alone calculate the marginal benefits and costs of those options and, based on those calculations, decide what action to take to maximize utility.

Instead of maximizing, Simon argued that people satisfice – a word that combines suffice and satisfy. Simon argued that real decision making is hermeneutic in nature. People engage in ordered search routines through a strictly limited set of alternatives. They make choices based on heuristics or tools that reduce uncertainty and the costs associated with gathering all of the necessary information for each decision. For example, they match problems to existing problems with known solutions and then adapt the old solutions to the new problems. Simon theorized that decision makers are confronted with various data pertaining to their actual environment. Given their perception of that environment and their beliefs about how the world works, they sort out their goals and objectives and decide in such a way as to satisfy their most salient objectives.

While Simon applied this pattern of satisficing to human behavior in general and administration in particular, he noted that it was especially applicable to budgetary decision making. Indeed, Simon's experience within a municipal finance office contributed directly to his formulation of bounded rationality (Simon, 1991). In budgeting, the details of the budget and the budgetary process, as well as the external and policy environment facing budgeters, not to mention the problem of successfully matching means to ends, are incredibly complicated. Simon noted that budgetary decision makers reduced uncertainty and information costs by basing expenditure and taxing decisions on last year's budget decisions.[2] However, Simon did not concentrate his efforts on developing an applied model of bounded rationality to the problem of budgeting but rather his students would go on to develop these models.

Charles Lindblom laid the second cornerstone of the organizational process model: the incrementalist approach to policy-makers and administration. Lindblom argued in "Policy Analysis" (1958) and in his seminal piece "The Science of Muddling Through" (1959) that policy-makers and administrators must confront the complexity of the real world. In this world, conflicting values and goals abound, policy alternatives are in short supply, and predictive theories of the outcomes of policy alternatives are often absent or unsatisfactory. Consequently, Lindblom, like Simon, insisted that optimization is impossible. Instead, he argued that policy-makers concentrate on only one or two goals at a time and consider only those alternatives that come to mind as a result of past experiences with the policy at hand. Because the reference point for decision making is always some decision made in the past, the outcome of local search, combined with a propensity to limit bargaining and enforcement costs, are small or incremental changes in policy.

Lindblom tended to use the terms policy maker and administrator interchangeably.[3] As John P. Crecine (1969) points out,

> Lindblom's administrator is a man with limited knowledge, limited information, and limited cognitive ability, making a policy choice in an uncertain world by "drastically" simplifying the problem and making marginal adjustments in past "successful" policies to formulate current policies. Lindblom goes on to argue that, in the face of uncertainty, limited information (and the "cost" of information), and limited computational ability (and the "cost" of expanding it), that incremental change in marginal adjustments of "proven" policies is probably a very rational way to reach policy decision (p. 11).

Although Lindblom's ideas about decision making are similar to Simon's (and, as we shall see, Aaron Wildavsky's as well), Lindblom focused more on outcomes than on individual decisions. This distinction is important because some later organizational process scholars (see Padgett, 1980a) denied that there was any fundamental difference in their approaches – that is not in fact the case – although perhaps this confusion is one of the sources of the present quandary over the organizational process model.

Lindblom's work continues to be influential, although more so among political scientists and public administrationists than among his fellow economists. However, incrementalism (Lindblom's as well as Wildavsky's) was subjected to a plethora of criticism during the 1970s (Wanat, 1974; LeLoup, 1977, 1978a, b). Later organizational process scholars, especially John D. Padgett (Jones, True & Baumgartner, 1998; Padgett, 1980a, 1981), would claim that incrementalism was not well formulated, that it was a metaphor rather than a fully-fledged behavioral model.

Aaron Wildavsky's main contribution to the development of the organizational process model can be found in his *The Politics of the Budgetary Process* (1964). *The Politics of the Budgetary Process* is a truly path-breaking work. It is easily the most frequently cited monograph in the history of political science. It almost single-handedly moved budgeting from the periphery of the discipline, where mundane administrative matters were relegated, to its core. Moreover, its explanation of government spending as the product of a set of games involving strategy and skill (e.g. ubiquitous and contingent strategies) and played by self-interested actors performing highly stylized roles (e.g. spender, cutter, guardian, etc.) remains fundamental to the way in which political scientists understand the budgetary process.

However, many political scientists seem to be unaware that Wildavsky actually adumbrated two theories of budgeting in *The Politics of the Budgetary Process*. Those who are aware of his second theory of budgeting are often also profoundly hostile toward it and its implications. Wildavsky's second theory was a working model of Simon's bounded rationality put into action. From the standpoint of this survey, however, this second model, the one in which Wildavsky showed the relevance of incrementalism to budgeting, is the more important of his two theories.

Furthermore, Wildavsky downplayed his first theory in *The Politics of the Budgetary Process*, where he observes that strategy and skill matter only if they influence outcomes. However, he stressed that spending outcomes are highly stable – they do not vary much over time, at least not in ways that can be better explained by economic rather than political variables. Consequently, Wildavsky concluded that, because the strategic approach to budgeting does not predict anything very interesting, it might be largely irrelevant as positive theory! Budget battles preoccupy the players. They generate a lot of sound and fury, but their significance overall is a lot more apparent than real.

Instead, Wildavsky concluded that "budgeting is experiential . . . budgeting is simplified, budgeting officials 'satisfice' . . . budgeting is incremental" (Padgett, 1980a, p. 371). Consequently, he asserted that outlay levels could, for the most part, be predicted using fairly simple models. He also asserted that while the incremental model might not predict major or non-incremental shifts in spending behavior, neither could the alternative explanation of government spending – as a product of games of strategy and skill played by self-interested actors performing highly stylized roles – although he insisted that that model had other uses.

FORMALIZATION OF THE ORGANIZATIONAL PROCESS MODEL

In collaboration with Otto Davis and Murray Dempster, Wildavsky later tested the assertion that outlay levels could be predicted using fairly simple models by estimating a set of simple incremental models of the appropriations process in the United States. Their formal model of incrementalism applied to budgeting was first presented in a 1966 article in the *American Political Science Review* (APSR) entitled "A Theory of the Budgetary Process" (Davis, Dempster & Wildavsky, 1966). In this article, Davis, Dempster and Wildavsky modeled bureau requests and congressional appropriations decisions. Refinements of this initial effort were presented in a series of articles published during the 1970s (Davis, Dempster & Wildavsky, 1971, 1974). The APSR article was more significant and influential in shaping later organizational-process research than any of its successors, however. It was also a major innovation for political science[4] in that it developed a mathematical model of the thick textured interpretation of the processes Wildavsky described in *The Politics of the Budgetary Process*.

Using simple linear models, Davis, Dempster, and Wildavsky demonstrated "the importance of 'aids to calculation' in substantially complex, yet institutionally stable problem environments (i.e. the budget problem). Institutional stability engendered stable mutual expectations, which markedly reduce the burden of calculations for the participants" (Padgett, 1980a, p. 355). Given institutional stability, decision makers could treat the decisions made within the process as reliable information. In the budget context, the core of reliable information is contained in last year's budget, or the "base," and this became the driving force in the elaboration of future decision rules governing bureau requests and congressional appropriations. Simple decision rules such as "grant to an agency some fixed mean percentage of that agency's base plus or minus some stochastic adjustment to take account of special circumstances" (Davis, Dempster & Wildavsky 1966, p. 532).

Of course, stability is contingent. External factors, what Davis, Dempster, and Wildavsky termed "shift points," could cause changes in the budget process or in budget outcomes. Once these factors were incorporated into the temporal pattern of decisions and the necessary adjustments made, Davis, Dempster, and Wildavsky theorized the process would return to stability. They further speculated that the normal decisions made in the budgetary process, as well as adjustments to special circumstances, reflected past learning by decision makers in the form of "standard operating procedures," which constrained the percentage adjustments made within the process (Davis, Dempster & Wildavsky, 1966, pp. 529–530).

Given this linear decision rule framework Davis, Dempster, and Wildavsky operationalized an array of simple regression equations. In 87% of the bureau cases Davis, Dempster and Wildavsky found that the empirically dominant incrementalist model of executive request behavior was:

$$\text{REQUEST}_t = \alpha_1 \text{ APPROPRIATION}_{t-1} + \varepsilon_t$$

Moreover, in 80% of the bureau cases, the empirically dominant model of congressional appropriation behavior was:

$$\text{APPROPRIATION}_t = \alpha_2 \text{ REQUEST}_t + \phi_t$$

While the work of Davis, Dempster, and Wildavsky has been widely criticized on two fronts: for its mathematical naiveté (Ripley, Franklin, Holmes & Morland, 1973; Wanat, 1974; Williamson, 1966) or for various alleged weaknesses that derive from incrementalism itself (Natchez & Bupp, 1973; Gist, 1974; LeLoup, 1978a, b), it remains an important milestone in budget theory and a keystone in the development of the organizational process model. If a good model is one that uses a little to explain a lot, their work constitutes a monumental effort. As Davis, Dempster, and Wildavsky explain, "this year's budget is based on last year's budget with special attention given to a narrow range of increases and decreases" in making decisions in the face of complexity.

The organizational process model is articulated in its clearest and most complete form in John P. Crecine's 1969 book *Governmental Problem Solving*. In *Governmental Problem Solving*, Crecine synthesized the work of Simon, Lindblom, and Wildavsky, as well as Richard Cyert and James March, which one might expect since both served on his dissertation committee. This original empirical study used computer simulation to examine budgeting in three major American cities: Detroit, Pittsburgh, and Cleveland. Based on interviews and studies of municipal budgeting dynamics, Crecine used computer simulations to recreate the environments and decision-making algorithms of the principal actors in municipal budgetary process: department heads, the mayor, and city council. Crecine's major assumption was that the budget had to be balanced against an exogenous revenue constraint, a stylized fact that is true by definition – operating budget deficits are strictly forbidden in local government. Crecine operationalized this assumption by requiring revenues to be equal to or greater than expenditures.

Crecine modeled three decision-making outcomes generated by the municipal budgetary process: (1) departmental requests; (2) the mayor's budget recommendations to the city council; and (3) the final council appropriation as approved by the city council (Crecine, 1969, p. 35). From the perspective of the department heads the budgetary problem can be viewed as developing a request that: (1) assures sufficient funds to carry on existing programs and to deal with existing exigencies; (2) is acceptable to the mayor's office, and; (3) provides a reasonable share of any overall budget increases to the department to enable it to attack new difficulties (if any) (Crecine, 1969, p. 39). Figure 1 shows Crecine's mapping of the process.

This process begins with the budget guidance letter from the mayor's office to the department heads that outlines the assumptions about revenues and expenditures they are to use when formulating their requests. The department heads base their requests on their previous budget plus an increment that reflects their interpretation of the budget guidance letter and its tone. Where department heads go beyond the scope of the letter (for example, a request of 2 to 5% over that outlined in the budget guidance letter), they must provide explicit and urgent justification for the proposed change. Therefore it is imperative that they accurately interpret the mayor's letter or their requests will face extraordinary budgetary scrutiny down the road in the process. In Crecine's experience and, therefore, in his simulation model, department heads closely followed the mayor's instructions. Crecine concluded that it would be unwise NOT to do so. Moreover, their failure to follow the mayor's instructions could jeopardize their relationship with the mayor and his or her staff for the sake of budget gains in a single year.

Crecine captured the intricacies of these relationships and the detailed processes in his three submodels representing the department, the mayor and the council. To illustrate how the submodels operate, Fig. 2[5] is Crecine's formal model of the general department. The department submodel begins with the budget letter and the budget forms received from the mayor that contain: (a) current appropriations for all account categories in the department; (b) current total appropriations; (c) the previous year's expenditures in various account categories; and (d) the estimate of allowable increases over current appropriations implied from the "tone" of the mayor's letter. Box 2 represents the trend of departmental appropriations – the direction and magnitude of recent changes in amounts of appropriations in departmental categories or perhaps more correctly shown in the figure "the memory of the department." Based on the information from boxes 1 and 2, the department formulates a reasonable request for funds in its existing account categories, using current appropriations as a base or reference point and adjusting the estimate according to whether there was an increase in appropriations last year and the difference between last year's expenditures and appropriations.

From this point, using reasonable requests calculated in box 3, a preliminary department total request is calculated. The next decision node, represented as box 5, is determining if the total department requests lie outside of the guidelines set by the mayor's office (which is implied from the tone of the mayor's budget

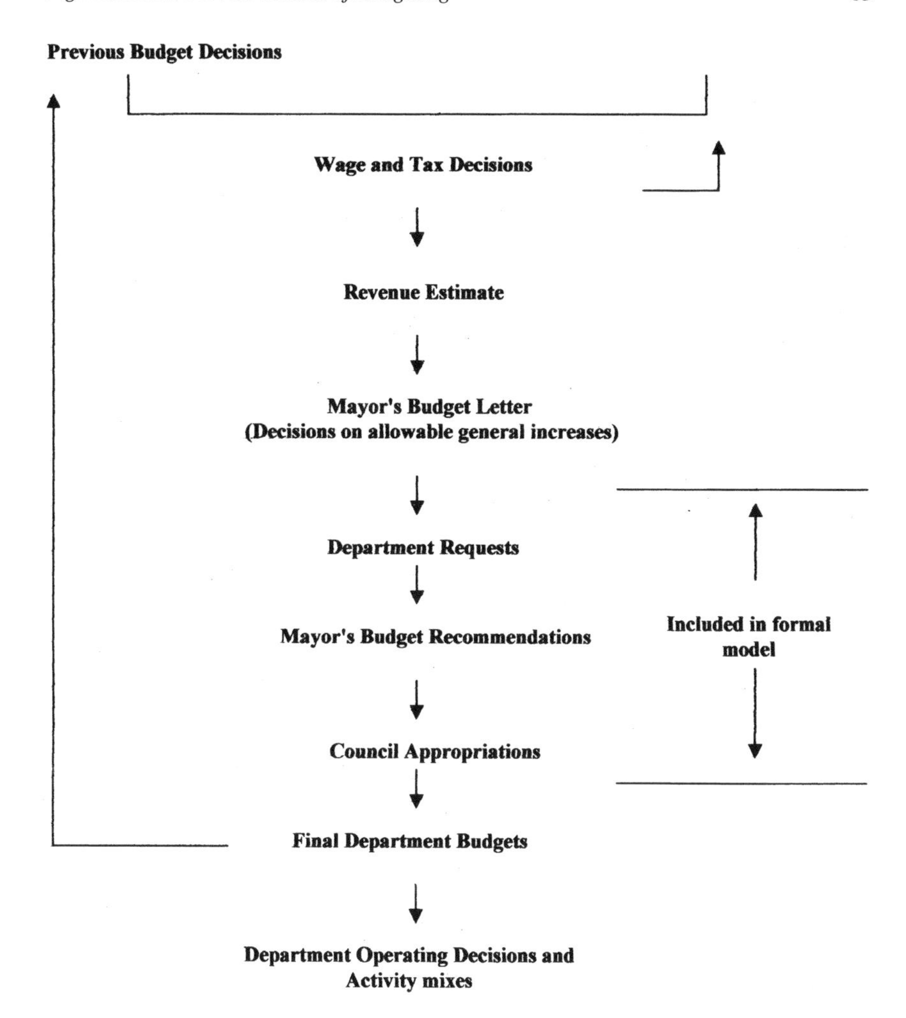

Fig. 1. Crecine's Hierarchy of Decisions and Model Overview.

letter). If no, the department checks to see if there are any increases in salary accounts over current appropriations. If yes, all department requests in all categories are adjusted so that any increase (proposed) over current appropriations is submitted as a supplemental request. The next step, then is to go to box 6 to check for salary increases. If there are increases in salaries, the department makes regular requests equal current appropriations and then places these increases in a supplemental budget request. If there are no increases or

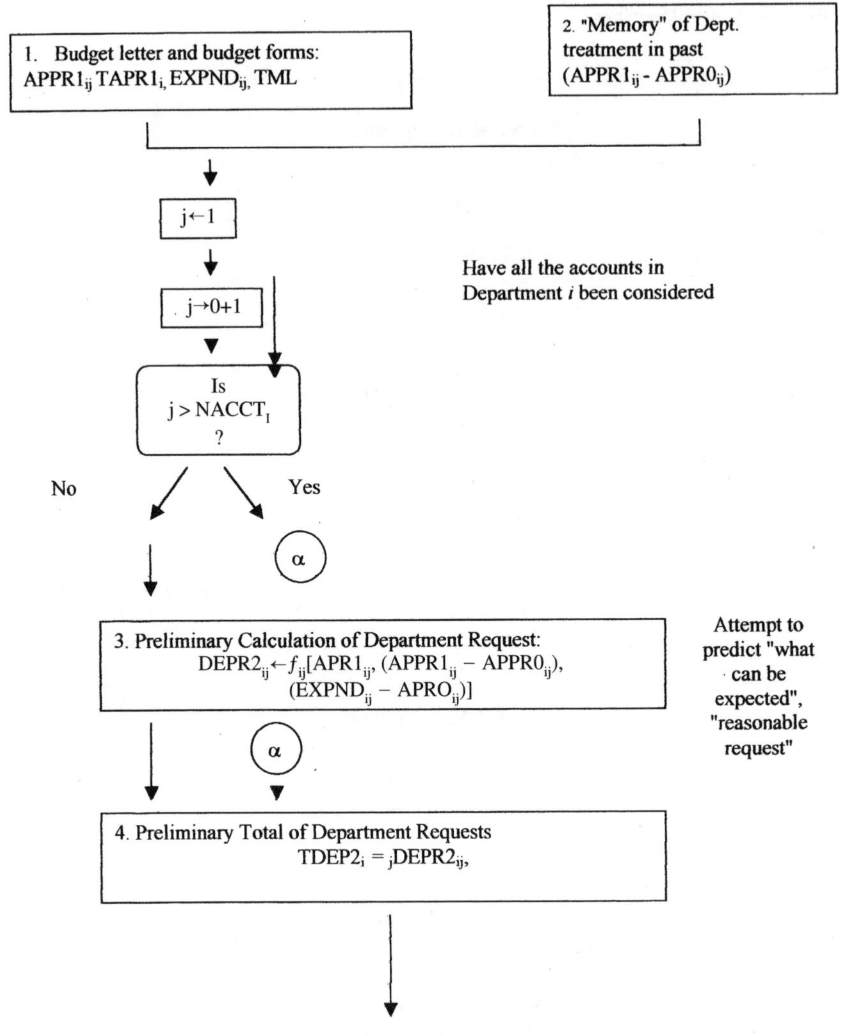

Fig. 2. Crecine's Department submodel.

after the supplemental requests have been developed, the next step is box 9, where the department calculates the total regular request. The final step is to send regular requests and the departmental total to the mayor's office along with any supplemental requests. It is from this point that the mayor's submodel captures the next phase of the budget process.

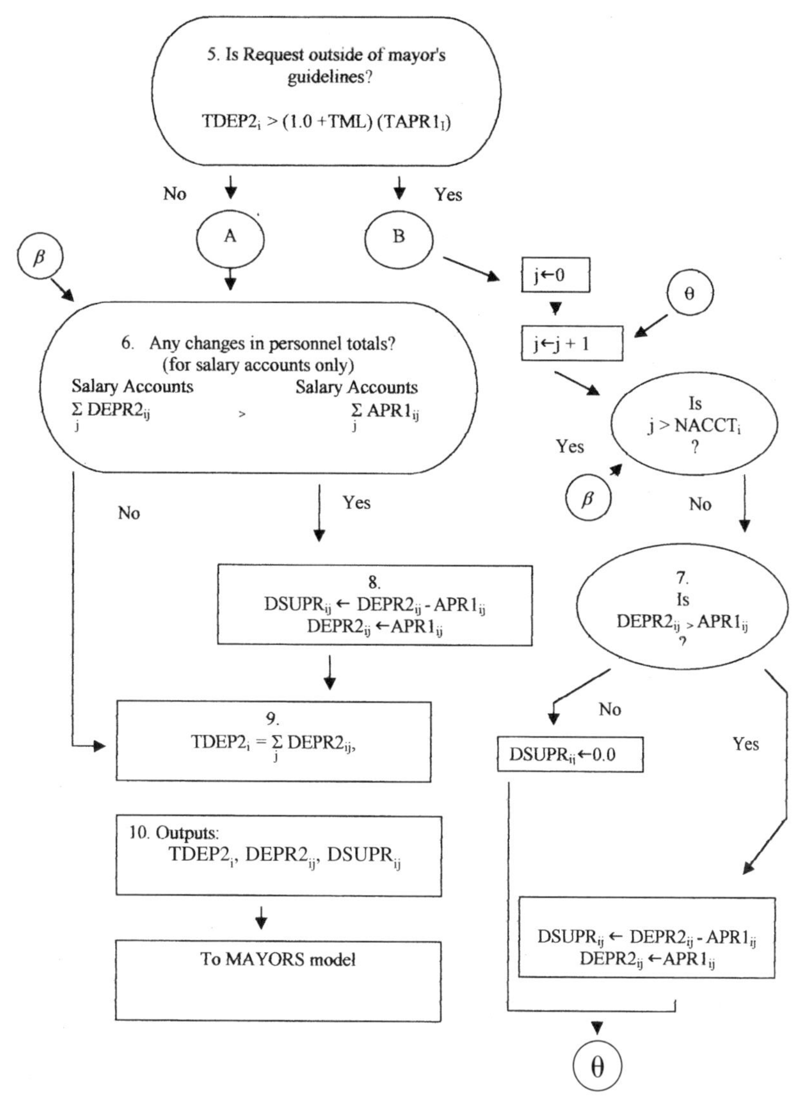

Fig. 2. Continued.

Crecine observed that none of the decision-makers in the budgetary process are primarily concerned with optimally allocating funds among functions to maximize the satisfaction of the citizenry; rather each looks at the process from

Table 1. Department Variable Key & Dictionary.

Variable	Definition
Model Inputs	
$APRO_{ij}$	Final appropriations, previous year
$APRR1_{ij}$	Final appropriations, current year
$EXPND_{ij}$	Expenditure total, previous year
$RMAY1_{ij}$	Mayor's budget recommendation, current year
TML	Tone of Mayor's budget letter for next budget year
$NACCT_{i}$	Number of account categories in the department
Accounting	
$TAPR1_{i}$	Total appropriations (all accounts) for department, current year
$TDEP2_{i}$	Total requests (all accounts) for department for next budget year
Parameters	
A_{ij}, B_{ij}, C_{ij}, D_{ij},	Empirically determined preliminary- calculations parameters (see Crecine 1967, pp. 61–66 and 259–263)
Intermediate Calculations and Model Outputs	
$DEPR2_{ij}$	Department budget request for next budget year
$DSUPR_{ij}$	Department supplemental budget request for next budget year
Subscripts	
i	Refers to a department
j	Refers to an account category within a department

All figures and this variable dictionary can be found on pages 55–59.

their own narrow perspective. While somewhat myopic, this division of attention can be explained in terms of simplifying a very complex problem by decomposing it into a set of manageable sub-problems (Crecine, 1969, p. 38). Even so, mayors confront very difficult, almost insoluble problems. In addition to balancing the budget, Crecine reminds us that mayors must also deal with myriad political and administrative problems. Mayors must maintain service levels, meet a payroll, and provide for mandatory increases in municipal wages and benefits, while simultaneously avoiding tax increases (especially rate increases on plant, equipment, and inventories since these can cause business to relocate). If, and only if, there are extra funds after solving these problems can mayors sponsor programs on their own political agendas. To manage these challenges, Crecine presumed that mayors would tend to rely on past precedent – which is also what he found.

From an organizational process perspective, the most interesting case is that of the city council. Crecine points out that decomposition of the budget problem drastically limits the role of council members. Because of the complexity and detail of the mayor's budget and the lack of council staff or other analytical resources, they are restricted to reviewing the mayor's recommendations, checking them for obvious errors or omissions, and then either approving or rejecting the mayor's budget in total (Crecine, 1969, p. 39).

Crecine points out that the budgetary process is framed by the mutual expectations of the principal actors and by their common understanding of each other's decision problems and rules. This division of attention not only reflects a municipality's hierarchy of authority and responsibility but it also implicitly allocates to each level of the hierarchy its own set of simplifying heuristics. Each actor can, therefore, anticipate the behavior of the other actors with reasonable confidence and independently proceed with their own decision-making.

In addition, Crecine claims that the principal actors rely on several other devices to further reduce the complexity of the budget process to more manageable dimensions. For example, they separate: (1) revenue forecasts from expenditure forecasts, both of which are independently estimated; (2) capital budgets from operating budgets, both of which are organized around a stable structure of administrative units (departments and bureaus) and account categories (salaries, supplies, expenses, equipment, etc.); and (3) the structure of municipal jobs into categories that are governed by uniform salary and benefit schedules. Again, decomposition defeats complexity by converting a few big problems into a series of smaller problems. Stable administrative and account structures simplify the task of recognizing and relying upon precedent. The presence of uniform wage policies avoids the potentially complex and entangling problem of dealing with employees on an individual basis (Crecine, 1969, p. 40).

Crecine insists that precedent is the key to budgetary decision making. In budgeting, precedent means taking last year's solution (current appropriations) and modifying that solution in the light of changes in available resources and current exigencies, given available remedies, to obtain this year's solution (Crecine, 1969, p. 41). Moreover, precedent implies that once an item is in the budget it will remain in the budget. Therefore, Crecine concludes, budgets evolve slowly as small changes are made in them from one year to the next, thereby making explicit the link between his organizational process models and Lindblom's and Wildavsky's incrementalism. To procedural considerations, Crecine adds a second reason for governance by precedent: the required openness of public decision making. Basing current decisions on precedent is completely defensible. Other bases of decision, he suggests, are much less so.

Several of Crecine's scientific innovations are noteworthy. In the first place, his focus on municipal expenditures was something of a novelty. Before Crecine, most rigorous analysts of state and local government – economists as well as political scientists and public administrationists – tended to direct their attention to the revenue side of the equation. The exceptions generally relied on ill-specified determinants models that randomly threw numbers of cross-sectional, explanatory variables at spending outcomes in the hopes that some would stick (as measured by standard statistical tests) (Borcherding, 1977).

His second significant innovation lay in the development of a distinctive methodology: computer simulation based on models developed from carefully constructed decision-making protocols. This innovation shaped almost all of the future work on organizational process models. Unfortunately, however, despite Crecine's remarkable success in explaining public expenditure outcomes, organizational scholars have largely abandoned this methodology.

His third innovation was to treat budget makers as "organizational decision makers and problem solvers who structure complex problems, generate alternatives, and make choices" rather than as discretionless agents of citizens' demands for locally provided goods and services (Crecine, 1969, p. 20) – the role that they implicitly play to this day in the bulk of the literature on pubic finance. Again, unfortunately, with the exception of a handful of his students, this innovation has largely been forgotten. In contrast, the parallel efforts of William Niskanen (1971, 1994) and his intellectual entourage to explain the spending rules that elected officials and bureaucrats are likely to follow given their incentives, have gained widespread attention and even adherence.[6] This is so despite their reliance on rather bizarre assumptions about bureaucratic tastes and opportunities and their rather poor explanatory power. Newer models that have been inspired by this intellectual tradition, but stressing the informational endowments of bureaucrats, the implicit and explicit contracts that link their actions to rewards, and their discretionary powers, often seem more plausible (Moe, 1990, 1989; Tabellini & Alesina, 1990; Cooley & Smith, 1989). They too have generally failed to yield successful empirical predictions beyond the ones for which they were expressly tailored.

Patrick D. Larkey was the next significant contributor to the development of the organizational process model. Larkey's award winning book[7] *Evaluating Public Programs* (1979) not only expanded the scope of the organizational process model but also further advanced the precision and utility of the model. In *Evaluating Public Programs*, Larkey examined the federal General Revenue Sharing (GRS) program, established by the Nixon administration which was in effect from 1972–1986. The distinguishing features of GRS that rendered it especially worthy of study included automatic eligibility (all jurisdictions – states, counties, cities, and Indian Tribes – automatically qualified for funding), formula-driven funding, an absence of programmatic restrictions, and minimal reporting requirements. Moreover, GRS did not require applications for funding, thereby offsetting the advantages enjoyed by the larger, usually more professional jurisdictions in obtaining federal grants.

At that time a number of economists and political scientists were concerned with intergovernmental transfers, some specifically with GRS. This effort had given rise to one of the most interesting and anomalous empirical results in public finance: the so-called flypaper effect, i.e. intergovernmental transfers tend to stick where they land (Oates, 1994; Ladd, 1994). That is: intergovernmental grants appear to produce far larger increases in the output of government services than predicted by the income and price elasticities of the basic demand model. Theory says that categorical grants should give rise to both income and substitution effects – they should permit the citizens of the receiving jurisdiction to increase their current savings and consumption by the amount of the grants and, to the extent that grants subsidized the consumption of specific services, they should cause citizens to increase their consumption of those services relative to others. In contrast, theory says that block grants, such as GRS, should have only minimal substitution effects and should, therefore, have little or no effect on the quantity of goods and services supplied by local jurisdictions. Thus the widespread finding that GRS increased government spending by nearly the full amount of the intergovernmental transfer seemed to call into question the most fundamental premise of public finance – that total demand for collectively provided goods and services is a function of the price of the goods, consumer income, and the size of the market. The obvious (although not the only possible) inference was that public officials were somehow short-circuiting the fiscal policy-making process so as to satisfy their preferences at the expense of those of the citizenry at large. This inference appeared to be an obvious invitation to organization-process theorists, with their emphasis on the role played by elected public officials in the determination of fiscal outcomes.

Larkey, however, had two objectives. He wanted to demonstrate the utility of the organizational process approach as a positive applied research methodology for policy evaluation as well as to evaluate the consequences of GRS. Consequently, Larkey's research design closely resembled Crecine's in a number of ways. Larkey relied on computer simulation based on simplified decision protocols. He relied on small samples of metropolitan areas (five cities in this case, although significant numbers of departmental accounts); and he relied on longitudinal data on the fiscal policy decisions and outcomes for each jurisdictional account as well as environmental data – demographics, revenue/liquidity constraints, etc. – to round out the hard data portion of the model. Like Crecine, Larkey grounded his model by interviewing the principal budget actors in each jurisdiction and, after he developed his protocols, revisited his informants to verify the accuracy of his simulation. The main thing that distinguished Larkey's work from Crecine's was its evaluative purpose. Crecine was concerned with accurately predicting spending. Larkey used Crecine's methods to build accurate counter-factual estimates of what spending would have been in the absence of GRS. He then compared his counterfactuals with actual spending to show the effect of GRS.

What Larkey found is that public-finance theory was generally right and that the econometric evidence for the flypaper effect was largely an artifact of errors in variables.[8] In other words, he found that GRS did not generally cause significant increases in the production/consumption of goods and services provided by municipalities. Instead, he found substantial misreporting in official reports. For example, outlays on public safety programs were substantially over-reported, while revenue displacement (tax reduction) was substantially under-reported. In the absence of liquidity constraints (Larkey used the term "fiscal pressure, " which conflates two different phenomena – revenue constraint and liquidity constraint – but in this instance it is clear from the context that he meant the latter, restrictions set by the market/rating bureaus on a municipality's capacity to borrow), the only significant increases in outlays that Larkey found involved shifts in the timing of cash flows not real changes in current consumption – the accumulation of surpluses (current assets), capital project and other investment spending (capital assets), providing for or reducing non-recurring obligations (liabilities), etc. That was not the case where fiscal policy-makers faced real liquidity constraints. Larkey found, as theory would suggest, the "greater the fiscal pressure," the more likely GRS funds were to be used to support or increase current operations. He also provided evidence to show that local governments are far more likely to be liquidity constrained than most finance scholars have assumed (e.g. Choate & Thompson, 1996; see, however, Fortune, 1998).

These findings are very important, but they have largely been overlooked. There are several possible reasons for this. Larkey's methodology, analytic categories and language are foreign to most public finance scholars; his use of the term fiscal pressure is just one case in point. Consequently, they have found it easy to ignore his analysis. But we are inclined to believe that the main reason for the failure of Larkey's findings to be incorporated into the public finance canon is that he just wasn't interested in the questions that were of greatest concern to economists and so failed to emphasize those findings.

Instead, Larkey chose to emphasize the utility of the organizational-process model as an alternative to rational choice and economic models of fiscal behavior. Larkey also emphasized the errors that can occur when evaluators uncritically rely on data supplied by subjects. Indeed, he insists that his analysis shows that those surveyed systematically bias their reports on program or policy consequences, perhaps, owing to the desire to present themselves to program sponsors in the most favorable possible manner.

Over the next ten years or so, Larkey contributed to the further development of the organizational process model, conducted empirical work testing the model and demonstrating its utility (Larkey & Sproull, 1981; Larkey & Smith, 1989; Larkey, 1981) and defended this approach as an alternative to rational choice models of fiscal policy-makers (Larkey, 1979).

DIFFERENTIATION OF THE ORGANIZATIONAL PROCESS MODEL

John Padgett conducted the next wave of distinctive work on the organizational process model. This work appeared in a series of articles in a diverse set of academic journals during the early 1980s. The first article was in the *American Political Science Review* in 1980. It was entitled "Bounded Rationality in Budgetary Research." In this article Padgett traced the theory of process incrementalism as adumbrated by Davis, Dempster and Wildavsky, applied it to budgeting and proposed an alternative model, which he called the theory of serial judgments. Padgett's theory of serial judgments has several elements in common with process incrementalism; e.g. it presumes "bounded rationality." But it can be differentiated from earlier incremental models by two distinctive features: (1) sequential search through an ordered set of discrete budgetary alternatives; and (2) non-deterministic final selection based upon the slightly ambiguous application of "informed judgment" (Padgett, 1980, p. 357).

In a concise survey of the development of the organizational-process model, Padgett credited three sources of bounded-rationality research with shaping his theory of serial judgements. The ultimate source of his theory was Simon's concept of "satisficing." As Padgett explained, his serial judgment theory, like Simon's, presumes that only a limited number of alternatives will be considered. But Padgett further postulated that the first acceptable alternative encountered would be chosen. In this instance, he referenced search models from the consumer-choice/information-processing literature. While Simon's notions of bounded rationality influenced this literature, its empirical formulations were more precisely specified. Finally, Padgett credited both Wildavsky and Crecine with directing his attention to the budgetary process and for incorporating complexity and hierarchy into the formulation of the budget problem.

Padgett's serial judgment theory is by far the most elegant specification of the organizational-process model to date. In his articles, Padgett built directly upon Crecine's approach to complex, adaptive organizational behavior, focusing on the supremely complex federal government, specifically the Office of Management and Budget (OMB), although Padgett eschewed the use of computer simulation to map out patterns of behavior and test his hypotheses. Consequently, Padgett "demonstrates how fiscal and political constraints from the presidential level (ceilings on departmental totals) can be filtered through the standard operating procedures of the OMB – cutting alternative amounts from different programs – and through the attention focusing patterns of OMB and the departments" (Bendor, 1990, p. 407).

Padgett's (1981) article in the *American Journal of Sociology* entitled, "Hierarchy and Ecological Control in Federal Budgetary Decision-making" set out to smooth the difference between the verbal description of budgeting shaped by bounded rationality and provide a better mathematical representation of OMB administrator's actual behavior (Bendor, 1990).

Padgett's model is best summarized by Jonathan Bendor's (1990) "Formal Models of Bureaucracy: A Review." Bendor states:

> In this model, the search of OMB examiners is guided by an overall departmental constraint, B, fixed by their superiors. Program submissions at this stage total A, so the examiners must find $\Gamma = A - B$ worth of cuts. In every round there is a fixed probability, θ_i, that an examiner's attention will focus on program *i*. This attention probability can vary across the *n* programs; of course $\sum_{i=1}^{n} \Theta_i = 1$
>
> Given that an examiner has focused on the *i*th program, he generates a limited number of salient cut alternatives, anchoring or starting his search at zero cut. (It is assumed that the magnitude of the cut alternatives is distributed exponentially; the parameters of the exponential vary across programs.) The examiner's task is completed when all the cuts he recommends sum to or exceed Γ. Because his attention across programs is governed by probabilistic process and because he generates alternatives probabilistically, the number of cutting rounds needed to reach the target Γ is a random variable as well (Bendor, 1990, p. 408).

This model was explicitly goal directed but it was not an optimization model but rather the model accurately mapped the span of bounded rationality for the OMB examiner (Bendor, 1990). Padgett's model solidified the weakest link in the organizational process model – a robust and sophisticated mathematical representation of bounded rationality.

DIFFUSION OF THE ORGANIZATIONAL PROCESS MODEL

The early 1980s marked the end of the formal development of the organizational process model. After 1986, Simon, Wildavsky, Lindblom, Crecine, Larkey and Padgett published only one other major article utilizing this approach.[9] The sudden disappearance of organizational process articles from research journals occurred at the worst possible time for the persistence of the model. Economic imperialism (of the academic variety) was marching from triumph to triumph. The application of economic logic – methodological individualism and rational, self-interested decision making – to questions and issues that have traditionally been the concern of political scientists and public administrationists was one of the great triumphs of modern social science. Moreover, economics' triumphs in this area were not confined to positive public choice – increasingly its logic was brought to bear on normative questions about public spending and even on questions of organizational behavior and processes.

Oliver Williamson, Gary Becker, Jon Elster, Douglass North and Vernon Smith took these topics into the mainstream of economic research. In so doing, they largely overlooked the organization process theorists, although Williamson and North genuflected to Simon and bounded rationality.

Moreover, Padgett left the field in disarray. His research had destroyed the coherence given to the organizational process model by the Carnegie-Mellon researchers. But his work failed to appeal to future researchers (see for examples, Zucker, 1987; Bendor, 1988; Masuch & Lapotin, 1989; Mandell, 1989) and even he did not persist with it.

Nevertheless, a few articles were published during the latter half of the 1980s and early 1990s that had the look and the feel of the organizational process model. The authors of these articles generally acknowledged an intellectual debt to organizational process theorists, but most did not emphasize it.

The work of Mark Kamlet and David Mowery is perhaps the most interesting case in point. Both were Crecine's students and worked with him to apply his version of the organizational process model to Federal budgeting (Crecine, 1975; Kamlet, Mowery & Crecine, 1980). They still contribute to the budget literature (Hahm, Kamlet & Mowery, 1997). Moreover, the analytical approach taken in their budget studies – rigorous statistical methods and time-series data on spending used to investigate the consequences of institutions (the Congressional Budget Act, Gramm-Rudman-Hollings, etc.) and political regimes (Reagan Administration) – reflects their academic provenance (Fischer & Kamlet, 1984; Hahm, Kamlet & Mowery, 1992, 1995a, b; 1996; Kamlet, 1983; Kamlet & Mowery, 1984, 1985, 1987; Su, Kamlet & Mowery,1987, 1993). Like Larkey, they build auto-regressive forecasting models to establish counterfactuals that they use to evaluate the budgetary consequences of political and policy variables. In every case their empirical models contain a substantial inertial component – last year's budget predicts this year's. This is a very impressive body of work. Arguably, it links public choice to the organizational process model. However, despite empirical weight of this inertial component, the organizational process justification for using past budgets to explain current budgets has largely disappeared from their work. Instead, they explain their models in terms of standard macroeconomic and public choice assumptions (see Thompson, 1997).

Two articles by Marc Choate and Fred Thompson (1988, 1990) sought to link organization process theorizing to the new economics of organization by treating decision makers in the budget process as agents in a principal-agent framework. In so doing, they recovered Larkey and Crecine's administrators, but they also relied on Padgett's sophisticated behavioral analysis to map the institutional rules that influence administrators' actions, thereby linking the scattered theories in the field to each other.[10]

Finally, in their highly influential monograph, *Agendas and Instability in American Politics* (1993), Frank Baumgartner and Bryan Jones argued that policy evolution is inherently incremental, even stable – most of the time. Episodically, the accumulation of causes, however, produces fundamental and widespread policy shifts. Baumgartner and Jones borrowed paleontologist Steven Jay Gould's notion of punctuated equilibrium to describe this evolutionary process. And, although Baumgartner and Jones were not specifically concerned with public budgeting in *Agendas and Instability of American Politics*, their methods and theory closely resembled those of the organizational process theorists. Their theory of policy evolution derives from the same cognitive limits as bounded rationality. Like Crecine, Baumgartner and Jones also emphasized organizational constraints on effective information processing, learning, and decision-making. Consequently, their analysis restored some of organizational incrementalism's tarnished intellectual luster. Moreover, the resemblance of their work to that of the organizational process theorists increased when James True, with his interest in budgeting, joined the research team of Baumgartner and Jones.

Baumgartner, Jones, and True have subsequently written a series of articles in which they apply their punctuated equilibrium model to the federal budget. In the first article of this series, "The Shape of Change: Punctuation and Stability in U.S. Budgeting 1947–94" (1996), Baumgartner, Jones, and True develop an agenda-based model of budgeting that is consistent with maximization models as well as the bounded rationality approach of decision-making. They find that budgetary outcomes are leptokurtotic and not Gaussian which supports their theory of patterns of stability interrupted by occasional major punctuations. Second, they find that budget changes are drawn from a Paretian probability distribution which they believe is consistent with Padgett's serial judgement model. This research has great potential to enliven the organizational process model but at this point in time Baumgartner, Jones, and True have only an elegant model that richly describes past budgetary behavior. However the challenge remains to develop this model so that it can accurately predict periods of stability and punctuations in the budgetary process.

The authors' aims in the second article of the series, "Does Incrementalism Stem from Political Consensus or From Institutional Gridlock?" (1998), are more modest. There, Baumgartner, Jones, and True investigate Wildavsky's account of the relationship between legislative comity and incrementalism (1992). Wildavsky had made the counter-intuitive claim that consensus produces incremental budgets and, conversely, that dissensus produces larger, more rapid spending changes. His critics argued the reverse, that dissensus produces gridlock; Baumgartner, Jones, and True used OLS regression to estimate annual changes in budget authority by subfunction for the period 1947 through 1995[11] to answer three questions: (1) Are budgets becoming more incremental (operationalized in terms of budget volatility)?; (2) Does dissensus (operationalized in terms of divided government, the percentage of bills vetoed by the President, and the polarization of the congressional parties) decrease volatility?; and (3) Is incrementalism caused by consensus or gridlock? They found that budget volatility is, in fact, decreasing, but that congressional dissensus increases budget volatility. They concluded that volatility "indicates dissensus and that budgeting was more volatile and probably less consensual in the past than in the supposedly rancorous present." This finding is significant because it apparently contradicts the predictions of incrementalism's many critics. In this article, Baumgartner, Jones, and True explicitly call attention to the organizational process theorists, even if they do not follow directly in their footsteps.

In the third article in this series, "Policy Punctuations: U.S. Budget Authority, 1947–1995" (1998), Baumgartner, Jones, and True tested their theory of punctuated equilibrium using a meticulously restored data set covering the evolution of obligational authority for every subfunction of the federal budget for the entire postwar period. Their results showed that federal spending is characterized by great stability most of the time. However, they also found that overall patterns of spending have been punctuated by two large spending shifts. According to Baumgartner, Jones, and True, these punctuations divide federal fiscal history into three distinct periods: postwar adjustment, from 1947 through 1955; robust growth, from 1956 through 1974, and restraint, from 1976 to the present. The conclusion they draw is that "punctuations occur, not just in some programs or subsystems, but also throughout government." They also claim that their model outperforms the three obvious rival budget drivers: economic change, partisan change, and shifts in public opinion. Once again, organizational process models outperform alternative economic formulations and also strictly political formulations.

Finally, B. Dan Wood, a well-known political scientist at Texas A&M University, has written a paper that fits loosely within the budget theory tradition outlined here – "The Balanced Budget Norm: Modeling Variations from 1904–1996." This paper evaluates the disequilibrium between federal revenues and expenditures as an error correction process, using time varying parameters estimated with Flexible Least Squares modeling techniques. Wood's model suggests that the balanced budget norm is actually stronger now than in the era prior to the Great Depression. Also, he finds no evidence that partisanship or divided government affects the strength of the equilibrating force associated with balancing the budget (see also Hicks, 1984; Kiewiet & McCubbins, 1985a, b). Finally, like Kamlet and Mowery but unlike almost all other budget scholars who discount the effect of Gramm-Ruddman-Hollings, Wood's analysis shows that the strength of the equilibrating force associated with expenditures stabilized after this legislation.

CONCLUSION

> There is a lack of a budget theory that could correctly describe how it is that public funds are spent for one thing rather than another, or prescribe how they should be spent V. O. Key (1940).

It was almost 60 years ago that V.O. Key made this now-famous observation regarding the status of budget theory. Many students of budgeting believe that Key's frustrated, somewhat folksy observation still characterizes the field. This belief is difficult to understand given the advances made by our science, yet it is the case.

Perhaps the simplest explanation comes from Thomas Kuhn's (1962) vision of science in which paradigms emerge and evolve as a result of competitive pressure from their intellectual rivals. In this field, rational choice has emerged as the dominant model. Public choice is the positive version of this model, benefit cost analysis the normative version. Many students of budgeting, particularly those of an institutional, political, or organizational bent, but some economists as well, cannot embrace this model. There is, after all, no reason for scholars to take it on faith and the simple fact remains that the champion has never been compelled to defend its title.

This is unfortunate for two reasons. First, it means that many, perhaps most, scholars remain outside the dominant paradigm and will not be persuaded to do otherwise. Second, the dominant paradigm will tend to hypertrophy. Paradigms need serious challenge if they are to evolve satisfactorily. The problem is that most of those who reject the rational choice model have little more than "thick detail" to offer as an alternative. That won't suffice.

Organizational process models have not for some time now, if ever, been accounted serious challengers to the dominant paradigm. We think that is a pity. In our opinion, this body of work remains the most interesting and rigorous alternative to the dominant paradigm. Moreover, advances in information technology have made its core methodology more accessible than ever before. We hope that contemporary budget researchers will reconsider the work of the organizational process theorists to see if it might be applied to questions of their own. Or, at the very least, we hope that they will engage us in a dialogue about the merits of the organizational process model.

NOTES

1. At first we thought that this unfamiliarity might be due to the term and not the thing itself. Organizational process theorists didn't call themselves organizational process theorists. Rather, the term and the classification come from Graham Alison (1971). But that is evidently not the case. One gets the same blank stare regardless of what you call the model or how you describe it.
2. The basic idea has been resurrected by economic historians under the rubric of "path dependence." See David, 1988.
3. The role of a policy maker and administrator with regard to the budget is a gray area, often dependent on the institution and institutional rules that affect the environment. See, for example, Choate & Thompson, 1988 and 1990.
4. Of course it should be noted that Toby Davis is a highly respected economist who was a founder of both the Public Choice Society and the Association for Public Policy and Management. Murray Dempster is a noted econometrician.
5. The figure is from pages 58 and 59 of *Governmental Problem Solving: A Computer Simulation of Municipal Budgeting*. The verbal description of the model in this paragraph is adapted from Crecine's verbal description of the process found in Fig. V-1a on page 55 of Crecine's book.
6. We think the popularity of these notions are due more to the attractiveness of structurally-induced equilibrium to political scientists with their interest in institutions and fascination with games of strategy and to economists who are uncomfortable with what would otherwise be a serious analytical anomaly – the evident over-supply of government services – than to their inherent utility.
7. This book was based on his 1975 dissertation, which also won the National Tax Association dissertation prize that year.
8. Industrial-organization economists have in recent years become cognizant of the degree to which discrepancies between economic concepts and accounting constructs have biased their empirical results (see Carlton & Perloff, 1998). This problem is far more serious where governmental entities are concerned, since budgetary accounting is typically on a cash basis and ignores the sources and uses of funds, especially changes in asset and liability stocks. Public-finance economists have, for the most part, not taken cognizance of its seriousness, however.
9. Larkey and Smith, 1989 *Policy Sciences* paper entitled "Bias in the formulation of local government budget problems." Wildavsky, of course, continued to write about

budgeting, but aside from the occasional commentary on the meaning of incrementalism, not directly within this tradition.

10. Their main finding was that on agency-theoretical grounds both public choice theorists and organizational process theorists were wrong in predicting conservative bias on the part of risk-averse budgeters, i.e. that they would tend to underestimate revenues and overestimate outlays. Instead, Choate and Thompson demonstrated that unbiased forecasting was strictly incentive compatible. At the same time, Cassidy, Kamlet, and Nagin (1989) independently advanced persuasive empirical analysis showing that there is "no systematic relationship between revenue forecast errors and state political and institutional factors . . . cast[ing] substantial doubt on the prevailing belief as found in the literature, that state revenue forecasts have a pronounced conservative bias."

11. For this section the years refer to fiscal years.

REFERENCES

Allison, G. T. (1971). *Essence of Decision; Explaining the Cuban Missile Crisis*. Boston: Little Brown.

Baumgartner, F. R. & Jones, B. D. (1993). *Agendas and Instability in American Politics*. Chicago: University of Chicago Press.

Bendor, J. (1988). Formal Models of Bureaucracy. *Brit. J. Polit. Sci.*, *18*, 353–395 part 3.

Bendor, J. (1990). Formal Models of Bureaucracy: A Review. In: N. Lynn & A. Wildavsky (Eds), *Public Administration: The State of the Discipline*. Chatham, NJ: Chatham House Publishers Inc.

Borcherding, T. E. (Ed.) (1977). *Budgets and Bureaucrats: The Sources of Government Growth*. Durham: Duke University Press.

Carlton, D. W., & Perloff, J. M. (1998). *Modern Industrial Organization*. Glenview, IL: Scott, Foresman/Little Brown Higher Education.

Cassidy, G., Kamlet, M., & Nagin, D. (1989). An Empirical Examination of Bias in Revenue Forecasts by State Governments. *International Journal of Forecasting, 5*(3), 321.

Choate, G. M., & Thompson F. (1988). Budget Makers as Agents. *Public Choice*, *58*(1), 3.

Choate, G. M., & Thompson, F. (1990). *Biased Budget Forecasts, Journal of Economic Behavior & Organization*, *14*(3), 425–434.

Choate, G. M. & Thompson F. (1996). Debt and Taxes. In: G. J. Miller (Ed.), *Handbook of Debt Management* (pp. 129–139). New York: Marcel Dekker.

Cooley, T. F., & Smith B. D. (1989). Dynamic Coalition Formation and Equilibrium Policy Selection. *Journal of Monetary Policy*, *24*, 211.

Crecine, J. (1969). A. Wildavsky (Ed.), *Governmental Problem Solving*. American Politics Research Series. Chicago: Rand McNally & Company.

Crecine, J. P. (1975). *Fiscal and Organizational Determinants of the Size and Shape of the U.S. Defense Budget*. Washington D.C.: Commission on the Organization of the Government for the Conduct of Foreign Affairs.

Cyert, R., & March J. (1963). *A Behavioral Theory of the Firm*. Englewood Cliffs, NJ: Prentice Hall.

David, P. W. (1988). *Path-Dependence: Putting The Past Into The Future of Economics*. Palo Alto, CA: Stanford Institute for Mathematical Studies in the Social Sciences Economic Series.

Davis, O. A., Dempster, M. A. H., & Wildavsky, A. (1966). A Theory of Budget Process. *American Political Science Review*, *60*, 529–547.

Davis, O. A., Dempster, M. A. H., & Wildavsky, A. (1971). On the Process of Budgeting II: An Empirical Study of Congressional Appropriations. In: R. F. Byrne et al. (Eds), *Studies in Budgeting*. Amsterdam: North Holland.

Davis, O. A., Dempster, M. A. H., & Wildavsky, A. (1974). Towards a Predictive Theory of Government Expenditure: US Domestic Appropriations. *British Journal of Political Science*, *4*, 419–452.

Fischer, G. W., & Kamlet, M. S. (1984). Explaining Presidential Priorities: The Competing Aspirations Levels of Marcrobudgetary Decision Making. *American Political Science Review*, *78*, 356.

Fortune, P. (1998). Tax-Exempt Bonds Really Do Subsidize Municipal Capital! *National Tax Journal*, *51*, 43.

Gist, J. R. (1974). *Mandatory Expenditures and the Defense Sector: Theory of Budgetary Incrementalism*. Pacific Palisades, CA: Sage.

Hahm, S., Kamlet, M S., & Mowery D. C. (1992). The Influence of the Gramm-Rudman-Hollings Act on Federal Budgetary Outcomes, 1986–1989. *Journal of Policy Analysis and Management*, *11*, 207.

Hahm, S., Kamlet, M. S., & Mowery D. C. (1995a). Institutions Matter: Comparing Deficit Spending in the United States and Japan. *Journal of Public Administration Research and Theory*, *5*, 429.

Hahm, S., Kamlet, M. S., & Mowery, D. C. (1995b). Influences on Deficit Spending in Industrialized Democracies. *Journal of Public Policy*, *15*, 183.

Hahm, S., Kamlet, M. S., & Mowery, D. C. (1996a). Postwar Deficit Spending In The United States. *American Politics Quarterly*, *25*, 139.

Hahm, S., Kamlet, M. S., & Mowery, D. C. (1996b). The Political Economy of Deficit Spending in Nine Industrialized Parliamentary Democracies. *Comparative Political Studies*, *29*, 52.

Hicks, A. (1984). Elections, Keynes, Bureaucracy and Class – Explaining United States Budget Deficits, 1961–1978. *American Sociology Review*, *49*, 165–182.

Jones, B. D., Baumgartner, F. R., & True, J. L. (1996). *The Shape of Change: Punctuations and Stability in U.S. Budgeting, 1947–1994*. In: 54th Annual Meeting of The Midwest Political Science Association. Chicago.

Jones, B. D., Baumgartner, F. R., & True, J. L. (1998). Policy Punctuations: U.S. Budget Authority, 1947–1995. *The Journal of Politics*, *60*, 37.

Jones, B. D., True, J. L., & Baumgartner, F. R. (1998). Does Incrementalism Stem from Political Consensus from Institutional Gridlock? *American Journal of Political Science, 60*, 23.

Kamlet, M. S., & Mowery, D. C. (1984). Games Presidents Do and Do Not Play: Presidential Circumvention of the Executive Branch Budget Process. *Policy Sciences*, *16*, 303.

Kamlet, M. S., & Mowery, D. C. (1985). The First Decade of the Congressional Budget Act: Legislative Imitation and Adaptation in Budgeting. *Policy Sciences*, *18*, 313.

Kamlet, M. S., & Mowery, D. C. (1987). Influences on Executive and Congressional Budgetary Priorities 1955–1981. *American Political Science Review*, *81*, 155–178.

Kamlet, M. S., Mowery, D. C., & Su, T. T. (1988). Upsetting National Priorities: The Reagan Administration Budgetary Strategy. *American Political Science Review*, *82*, 1293–1307.

Kamlet, M. S. (1993). Modeling U.S. Budgetary and Fiscal Policy Outcomes: A Disaggregated, System wide Perspective. *American Journal of Political Science*, *37*, 213.

Key, V. O. (1940). The Lack of a Budgetary Theory. *American Political Science Review*, *34*, 1137–1140.

Kiewiet, D., & McCubbins, M. (1985a). Appropriations Decisions as a Bilateral Bargaining Game Between President and Congress. *Legislative Studies Quarterly*, *10*, 181–201.

Kiewiet, D., & McCubbins, M. (1985b). Congressional Appropriations and the Electoral Connection. *Journal of Politics*, *47*, 59–82.

Kuhn, T. S. (1962). The Structure of Scientific Revolutions. Chicago: University of Chicago Press.

Ladd, H. F. (1994). Measuring Disparities in the Fiscal Conditions of Local Governments. In: J. Anderson (Ed.), *Fiscal Equalization for State and Local Government Finance*. Westport, CT: Greenwood Press.

Larkey, P. D. (1979). *Evaluating Public Programs: The Impact of General Revenue Sharing on Municipal Government*. Princeton: Princeton University Press.

Larkey, P. D. (1981). Fiscal Reform and Government Efficiency: Hanging Tough. *Policy Sciences*, *13*, 381.

Larkey, P. D., & Smith, R. (1989). Bias in the Formulation of Local Government Budget Problems. *Policy Sciences*, *22*, 123–166.

Larkey, P. D., & Sproull, L. S. (1981). Models in Theory and Practice: Some Examples, Problems, and Prospects. *Policy Sciences*, *13*, 233–246.

LeLoup, L. (1977). *Budgetary Politics: Dollars, Deficits, Decisions*. Brunswick, OH: King's Court.

LeLoup, L. (1978a). The Myth of Incrementalism: Analytic Choices in Budgetary Theory. *Polity*, *10*, 488–509.

LeLoup, L. (1978b). Review of Aaron Wildavsky, Budgeting. *American Political Science Review*, *72*, 351–352.

Lindblom, C. (1959). The "Science" of Muddling Through. *Public Administration Review*, *19*, 79–88.

Mandell, M. B. (1989). Simulation Models and the Study of Information Utilization in Public Sector Decision-Making. *Knowledge*, *10*, 202–214.

Masuch, M., & Lapotin, P. (1989). Beyond Garbage Cans: An AI Model of Organizational Choice. *Administrative Science Quarterly*, *34*, 38–67.

Moe, T. M. (1990a). The Politics of Bureaucratic Structure. In: J. E. Chubb & P. E. Petersen (Eds), *Can the Government Govern?* Washington D.C.: Brookings Institution.

Moe, T. M. (1990b). The Politics of Structural Choice: Toward a Theory of Public Bureaucracy. In: O. Williamson (Ed.), *Organization Theory: From Chester Barnard to the Present and Beyond*. New York: Oxford University Press.

Mowery, D. C., Kamlet, M. S., & Crecine, J. P. (1980). Presidential Management of Budgetary and Fiscal Policymaking. *Political Science Quarterly*, *95*, 395.

Natchez, P. B., & Bupp, I. C. (1973). Policy and Priority in the Budgetary Process. *American Political Science Review*, *67*, 951.

Niskanen, W. (1971). *Bureaucracy and Representative Government*. Chicago: Aldine Press.

Niskanen, W. A. (1994). *Bureaucracy and Public Economics*. Brookfield, Vermont: Edward Elgar Publishing Company.

Oates, W. E. (1994). Federalism and Government Finance. In: J. Quigley & E. Smolensky (Eds), *Modern Public Finance*. Cambridge MA: Harvard University Press.

Padgett, J. F. (1980a). Bounded Rationality in Budgetary Research. *American Political Science Review*, *74*, 354–372.

Padgett, J. F. (1980b). Managing Garbage Can Hierarchies. *Administrative Science Quarterly*, *25*, 583–604.

Padgett, J. F. (1981). Hierarchy and Ecological Control in Federal Budgetary Decision Making. *American Journal of Sociology*, *87*, 75–128.

Ripley, R., Franklin, G., Holmes, W., & Morland, W. (1973). *Structure, Environment, and Policy Actions: Exploring a Model of Policy Making*. Beverly Hills, CA: Sage.

Simon, H. A. (1955). A Behavioral Model of Rational Choice. *Quarterly Journal of Economics, LXIX*, 99–118.

Simon, H. A. (1957a). *Models of Man*. New York: Wiley.

Simon, H. A. (1957b). *Administrative Behavior*. New York: Macmillan Company.

Simon, H. A. (1991). *Models of My Life*. New York: Basic Books.

Su, T. T., Kamlet, M. S., & Mowery, D. (1993). Modeling U.S. Budgetary and Fiscal Outcomes: A Disaggregated, System wide Perspective. *American Journal of Political Science, 37*, 213–245.

Tabellini, G., & Alesina, A. (1990). Voting on the Budget Deficit. *American Economic Review, 80*, 37.

Thompson, F. (1997). Public Choice. In: *International Encyclopedia of Public Policy and Administration*. Boulder, CO: Westview Press.

Tversky, D., Kahneman & Amos (1984). Choices, Values and Frames. *American Psychologist, 4*, 341–350.

Wanat, J. (1974). Bases of Budgetary Incrementalism. *American Political Science Review, 68*, 1221–1228.

Wildavsky, A. (1964). *The Politics of the Budgetary Process*. Boston: Little, Brown.

Wildavsky, A. (1992). *The New Politics of the Budgetary Process*. Harper Collins.

Williamson, O. E. (1966). A Rational Theory of the federal Budgetary Process. In: G. Tullock (Ed.), *Papers on Non-Market Decision Making*. Charlottesville, VA: Thomas Jefferson Center for Political Economy.

Zucker, L. G. (1987). Institutional Theories of Organization. *Annual Review of Sociology, 13*, 443–464.

5. THE MEDIAN VOTER MODEL IN PUBLIC BUDGETING RESEARCH

Paula S. Kearns and John R. Bartle

ABSTRACT

The median voter theorem is a useful construct for the analysis of government budgeting decisions in democratic systems. Both the microeconomic paradigm of individual choice and the American political tradition of voter sovereignty contribute to the appeal of the explanatory value of the median voter model. However the model as currently articulated has serious limitations that endanger its continued viability. This essay first reviews the theoretical construction of the median voter model, its applications to government budgeting, and the empirical support for the model. Then, the limitations of the model are explored and potential steps to address its shortcomings are highlighted. Finally, the potential for future research based on the median voter model is assessed.

INTRODUCTION

Economic theory, and more specifically public sector economics, is the basis for many contemporary studies in public budgeting and finance. This approach considers collective decisions the result of the aggregation of individual choices made by rational, utility-maximizing actors. One aspect of economic theory that is especially pertinent for the assessment of budgetary outcomes is the median voter model. This model is derived from what is often referred to as the median

Evolving Theories of Public Budgeting, Volume 6, pages 83–100.

ISBN: 0-7623-0790-0

voter theorem, which states that under certain assumptions, the preferences of the median voter will be decisive. The median voter model as applied to government fiscal choice specifies an optimization model where individual utility for public and private goods is maximized subject to the combined public and private resource constraint. In this context the median voter hypothesis is that a single individual – the median voter – will be decisive in budgetary decisions, and the resulting public budget is then a reflection of his or her preferences for public goods. The median voter hypothesis was first posited by Bowen (1943), and the median voter model was extended to government expenditures in significant detail by Barr and Davis (1966), Henderson (1968), Borcherding and Deacon (1972), Bergstrom and Goodman (1973), Rubinfeld (1977) and Inman (1978, 1979).

The model is well-suited to empirical statistical analysis, because of the availability of data and ease of interpretation. Applications of this approach to the study of public budgeting model expenditure outcomes as if they had been determined by the median voter. This greatly simplifies aggregate analysis of expenditures because only characteristics of the median voter's income, tax price, and preferences are needed. The median voter model has applications to local, state, and federal expenditures, and has been generalized to international trade coalitions. Empirical evidence is somewhat supportive of the model, especially in certain cases at the local level. The limitations of the model are important, yet it persists as a common model for statistical analysis of budgetary outcomes and as a foundation for extended analysis. However, because the median voter model as currently articulated is devoid of political, institutional, and historical variables which are known to be important, the model is problematic.

This chapter discusses the median voter model as it applies to the study of public budgeting. The literature drawing from this perspective is voluminous. This chapter does not attempt to serve as a comprehensive literature review, but rather an assessment of the utility of this model for future research in public policy and administration. The next section presents the theoretical model's assumptions and implications, followed by a review of the most important literature. Then the model is critically evaluated, followed by a conclusion assessing the future research potential of the model.

THEORETICAL MODEL

Assumptions

Because governments are involved in the provision of collective goods, they often lack market pricing mechanisms for the efficient allocation of goods and services. In a democracy, individual votes provide a convenient proxy for consumer choice. Only unanimous votes guarantee that government decisions will be efficient, in the Pareto sense.[1] Even very small groups sometimes have difficulty reaching unanimous decisions in households, neighborhoods, and faculty meetings. Unanimity is often an unrealistic criterion for public decision making, so majority rule votes are used for making many collective decisions. Although some decisions (e.g., amending constitutions, extending debt limits, over-riding vetoes) require super-majorities, many recurring decisions (e.g. electing representatives, passing referenda) are made by simple majority rule. The alternative which can garner the support of one more than 50% of the voters will be the successful alternative. The voter that puts the plurality over the 50% mark to a majority is identified as the median voter, and is decisive in majority rule votes.

The median voter model assesses the outcome of majority-rule decisions, and is based on just a few powerful, simplifying assumptions: single-peaked preferences, unidimensional choices, and the absence of agenda setting. After examining each of these assumptions, the implications of the model will be assessed, and the empirical findings evaluated.

It should be noted here that a distinction is often made between the median voter *model* and the median voter *hypothesis*. The former is a construct for analyzing the demand of voters for government outputs. It may be paired with a model of government output supply. The latter is, as Turnbull and Mitias (1999) write, "the much stronger proposition that the community's fiscal behavior in equilibrium reflects the outcomes that are most preferred by the median voter" (p. 119). This distinction between these two terms is maintained in this chapter.

Single-Peaked Preferences.

The peak of any individual's preferences is that point compared with which all other neighboring points are less preferred, or lower. Peaks can be local or global, like the optimal points of other mathematical functions. Single-peaked preferences[2] means that the farther an individual moves from his or her most preferred point, the lower the benefit or utility. For single-peaked preferences, the global peak is also the only local peak in the preference function.

Figure 1 provides an illustration of single-peaked and non-single-peaked preferences. Individual A most prefers high spending, then moderate spending, and least prefers low spending. Individual B most prefers low spending, then high spending, and least prefers moderate spending. Individual C most prefers moderate spending, then low spending, and least prefers high spending. Individuals A and C exhibit single-peaked preferences – the farther they move

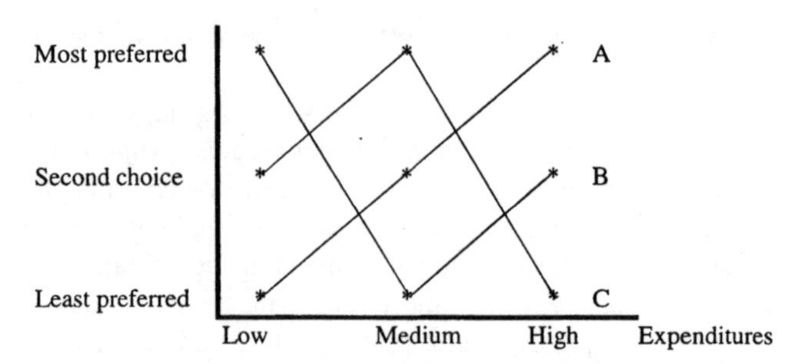

Fig. 1. Single-Peaked and Non-Single-Peaked Preferences.

from their most preferred spending level, the less their utility. Individual B exhibits non-single-peaked preferences – his utility decreases by moving from low spending to moderate spending, but increases somewhat when spending is high.

Unidimensional Choice

Another restrictive, yet simplifying, assumption of the median voter model is that of unidimensional choice. This refers to decisions that involve only a single choice parameter; expenditures and ideology are common choices in empirical work. This assumption suggests that the model would be most well-suited to the analysis of referenda, the study of direct democracies, or the study of committee decisions.

Absence of Agenda-Setting

Originally, agenda-setting referred to the presentation of alternatives in some order which would influence the outcome. Agenda-setting has come to include all sorts of strategic behavior: vote-ordering, strategic voting, sophisticated voting, vote-trading, or any calculated effort to deliberately subvert the majority rule system. At its simplest, the absence of agenda setting means that all alternatives will be considered and in no particular order.

Who is the Median Voter?

Finally, who is the median voter if he or she is actually decisive? Even in a referendum or a town hall meeting it would be very difficult to identify the median voter. A seemingly reasonable simplification commonly used in much of the literature focusing on government spending is that the median income resident is the median voter. This has the further advantage that data on median income is regularly reported by the U.S. Census Bureau, allowing the analyst to use income as a variable, and therefore to estimate the income elasticity of public spending.

Implications

If these assumptions are satisfied, then the outcome of a majority-rule decision will reflect the preferences of the median voter, who may further be assumed to be the median income individual. The median voter, as the name implies, is the voter whose preferences lie between those of other voters. In budgetary decisions, the median voter will prefer as much or more spending as half of the voters and as much or less spending as the other half of the voters. If voters are uniformly distributed, half of all other voters will prefer more government expenditure, half of all other voters will prefer less government expenditure. If voters are normally distributed, then the median voter will also be the modal voter, so his/her spending preferences will not be unique.

If preferences are discrete and unique, the median voter is easily identified. Consider, for example, seven voters with the following expenditure preferences: V_1 prefers $5, V_2 prefers $10, V_3 prefers $15, V_4 prefers $25, V_5 prefers $30, V_6 prefers $35, V_7 prefers $45. Majority voting on expenditures will result in $25, the preference of V_4. Three voters prefer more, and three voters prefer less. With a single voter associated with each expenditure possibility, most voters will be dissatisfied with the outcome.

Consider the possibility that groups of voters prefer particular expenditure amounts. Suppose 5% of voters prefer $5, 10% prefer $10, 20% prefer $15, 30% prefer $25, 15% prefer $30, 15% prefer $35, and 5% prefer $45. In this case, the median voter belongs to the group that prefers $25, because the fiftieth percentile is exceeded within that group. While 35% prefer less, and 35% prefer more, 30% of the voters are completely satisfied with the outcome. Even when the median voter is decisive, most voters are dissatisfied with the outcome.

One apparent normative appeal of the median voter model for budgeting research is that it is consistent with democratic notions of majority rule. However the median voter model is susceptible to Arrow's Impossibility Theorem if any of its restrictive assumptions are violated. Possible results include problems involving cycling, and intransitivity. Cycling is the potential for the vote between three or more alternatives to endlessly cycle between the alternatives, implying that the winning outcome is determined in part by the order in which the alternatives are considered on the agenda. Transitivity is the property that if outcome A is preferred to outcome B, and B to outcome C, then A is preferred to C. Both of these desirable characteristics of a voting system may be violated if the assumptions of the model are not met. Of course

Arrow's theorem proves that *all* voting systems are subject to certain problems, but it becomes clear that the normative attractiveness of the model is diminished when its assumptions do not hold.

Empirical Applications

Practically, the median voter model is well-suited for empirical work, despite its theoretical shortcomings. It allows for the estimation of demand functions for public goods based on readily available data from U.S. Census Bureau for many sub-national jurisdictions. This relative ease of data collection, backed with a coherent theoretical framework, makes this model very attractive for researchers seeking the path of least resistance.[3]

Suppose each member of a community received the amount of government service which she demanded. The demand function would be of the form:[4]

$$\ln G = B_1 + B_2 \ln Y + B_3 \ln P_G + B_4 \ln N + B_5 \ln Z + e \quad (1)$$

where G is government expenditure, Y is individual income, P_G is the individual tax price, N is population of the jurisdiction, and Z is a vector of taste variables. B_2 estimates the income elasticity of demand, B_3 the price elasticity of demand, and B_4 estimates g, a coefficient indicating the "publicness" of the good on an interval from zero (a pure public good) to one (a pure private good). A number of empirical papers using the median voter model demonstrate the usefulness of the model for several purposes. As Thompson writes, "[l]iterally hundreds of studies have demonstrated that family/per capita income, tax price, community size, and population served better explain cross-sectional variations in collectively supplied service levels than any other set of 'determinants' " (p. 1006).

Bergstrom and Goodman (1973) conducted a classic study based on the median voter model. It is typical of median voter models, and therefore instructive. They estimated three dependent variables: total expenditures, police expenditures, and recreation expenditures using the explanatory variables median income, tax price, population, population density, population change, employment-resident ratio, percent elder population, percent minority population, and percent owner-occupied housing. The income, price, and population variables correspond to those in Eq. (1). The remaining variables are proxies to represent the taste of the median voter. They found that price elasticities were negative and inelastic for all three dependent variables. Income elasticities are positive, as would be expected for normal goods. Demand for parks and recreation was more income elastic than demand for other services. This is consistent with intuition that police and total expenditures are driven by necessity, while parks and recreation are more of a luxury from the view of the median voter of a municipality.

The use of median income is reasonable, though not without controversy, as discussed below. Tax price attempts to measure the net cost to the median voter of the public service. Thus it varies with factors such as intergovernmental aid, federal income tax deductibility, the distribution of the local tax burden, the cost of producing local services, and the median voter's share of the tax base. This shows the power of the model, as it can predict changes in expenditure variables as a result of these factors affecting tax price. Population is straight-forward. Using variables such as employment-resident ratio, percent elder population, and percent minority population as proxies for the preferences of the median voter is clearly problematic. To be true to the model, this individual's preferences would be included in some way. Demographic and employment information can only be justified by empirical verification and convenience. Despite this weak justification, this approach is standard.

LITERATURE REVIEW

Some particularly important studies that allow for an evaluation of this model include Henderson (1968), Borcherding and Deacon (1972), Rubinfeld (1977), Inman (1978, 1979), Pommerehne and Frey (1976), Pommerehne (1978), Grosskopf and Hayes (1983), Baumgardner (1993), and Turnbull and Mitias (1999). Henderson (1968) articulated an explicit model of local finance where social welfare was maximized subject to a public-private budget constraint. His regression equations for the spending of all local governments within each county reported R^2 of 0.651 and 0.552 for equations with three independent variables. Borcherding and Deacon (1972) articulated a model which assumed that the public services produced by a representative democracy reflects the preferences of the median voter and are supplied efficiently by the government. They estimated price and income elasticities and the "publicness coefficient" for eight categories of public services.

Inman (1978) tested the model for 58 New York school districts and found the model to be a reasonably accurate fit with the data. More specifically, he found empirical support for the assumption that the person with the median income is the median voter. This is important because along with the other assumptions discussed above, it justifies the assertion that government spending levels will reflect the preferences of the person with the median income. This then justifies application of the familiar economic model of consumer behavior to the analysis of government spending decisions. This article therefore provides one of the critical empirical justifications for the model that is presented in the chapters in economic textbooks dealing with government fiscal behavior: the choice of the government in question is between two goods (a private good and a public good, or two public goods) and utility is maximized subject to the resource constraints. Further, interventions such as intergovernmental grants are modeled within this framework, generating a persuasive positive and normative model of government.

Inman (1978) concluded that,

> Using the median income voter as the pivotal decision maker therefore appears to be a reasonable first-order approximation to the process of local politics in this sample of single service governments. If so, we can model local fiscal choices as an individual utility maximizing problem and transfer to government behavior the usual theorems of consumer choice. Politics becomes economics . . . This is *not* to say we can now turn away from further empirical analyses of the process of local fiscal choice. It is important to know why a theory predicts well (pp. 59–60, emphasis original).

Despite Inman's qualifications and cautions, the median voter model has been extended to many other contexts, most notably general purpose governments. And while there are serious concerns with the generality and utility of this model, there is no question that, as Inman puts it, such models "gain in analytical simplicity without losing a great deal in predictive accuracy" (p. 60).

Inman (1979) extended the median voter model by articulating a general preference function for public and private goods, a cost structure equation for public goods, and individual and government budget constraints which yield a series of demand equations for public and private goods. An important advantage of this model is that it is flexible enough to allow the addition of any relevant independent variables to the analysis to isolate the predicted effect of this variable on the dependent variable(s). For example, Craig and Inman (1986) use this model to attempt to predict the impact of Reagan's "New Federalism" initiative on education and welfare spending.

However in his 1979 paper, Inman admits that the median voter model as articulated in its common form is too simple and restrictive. He has proposed several extentions. First, the development of a general equilibrium model making endogenous such variables as factor prices, grants, tax base, and population. Second, the incorporation of a simplified yet realistic bargaining model of local budgeting with two phases: expenditure determination and production. Third, with Craig (1986) a model of "structure-induced equilibrium" where institutional structural characteristics (rules for who is allowed to make proposals, their consideration, and amendment) and the preferences of political actors help determine fiscal outcomes. Finally, with Ingerberman (1988) a model incorporating compromise, institutions, and history.

Pommerehne (1978) assessed the expenditures of 111 Swiss cities, and found the strongest support for the median voter model in direct democracies, followed by representative democracies with referenda, and the least support in representative governments without referenda. Mueller (1989) writes that these findings suggest that representative government does not simply convey the preferences of the median voter. This then draws into question the key assumption of the model. Pommerehne and Frey (1976) test the model against the possibility that it is the *mean* voter who is decisive, and find some empirical evidence in support of the median voter model.

Grosskopf and Hayes (1983) find that the restrictions imposed by the median voter model's assumption of "as if" preference maximization by a single voter were not empirically verified from a sample of 132 Illinois municipalities. More specifically, the functional form imposed by the standard median voter model as applied to cross-sectional data did not perform as well as a more general "translog" (transcendental logarithmic) utility function that imposes fewer restrictions. The restrictions of the median voter model caused changes in signs and the magnitude of coefficients. They conclude,

> This result lends support to those who argue that public sector decision making cannot be modeled using the neoclassical approach. This result is consistent with the public choice, incrementalist, and organic approaches to public sector decision making. All of these approaches contend that the governmental decision making process is not strictly economic, but rather that the institutional structure affects outcomes either through the influence of the decision makers (public choice), the size of last year's budget (incrementalist), or the public will (organic approach) (pp. 211–212).

Turnbull and Mitias (1999), using a rigorous specification test of the median voter model versus a spending model using mean income and tax share variables, find that neither model does well. Further, the median voter model does worse than the alternative model. The median voter model does not explain U.S. state or county spending well, although it does reasonably well in explaining municipal spending. Similarly, Baumgardner (1993) finds that a political support maximization model that sees policy outcomes as the result of the pressures exerted by interest groups outperforms a median voter model in examining state Aid to Families with Dependent Children and Medicaid benefit and coverage rates. Thus the generalizability of the model is subject to question, especially in comparison to other models.

Several studies, such as Inman (1979), Bergstrom and Goodman (1973), Borcherding and Deacon (1972), and Mueller (1989) question the accuracy of empirical estimates of the "degree of publicness" parameter. Many estimates of this parameter are close to one, although this estimate is tentative. Such a value implies that congestion or other factors require facilities used in public goods production to grow at the same rate as population. That is, though the goods are provided publicly, they have the rivalry characteristics of private goods. This may be due to the lack of economies of scale in the cases examined. It does not imply that local publicly-provided goods are or should be private goods.

An anomaly produced by this model is the so-called "flypaper effect," which finds a significantly higher effect on local spending of intergovernmental aid compared to the effect of private income. This voluminous literature is better discussed elsewhere (see Hines & Thaler, 1995; Duncombe, 1996; and Bailey & Connolly, 1998). However because it consistently presents an empirical finding that does not square with theoretical prediction, it brings into question the veracity of the model.

Some general patterns have emerged in the empirical literature on the median voter model. Not surprisingly, it seems to do best the closer the situation is to the assumptions of unidimensional choice and direct democracy. As the government in question moves from a single-service district to a multi-purpose one, and as there are important mediating influences (such as interest groups and appointed or elected officials) between citizens and spending outcomes, the results of these various studies are less often empirically verified. Also, alternative models outperform the median voter model in several studies. So while the model is in some cases a reasonable representation of reality, in many others it is not, and often other models (including some non-economic models) have been shown to fit the data better.

On the positive side, the median voter model can be specified to explain either spending or revenues. It has been used to estimate the demand for education, corrections, health care, recreation, and infrastructure. Several such models have been tested and the empirical results from these models have shown them to have relatively high explanatory power, typically 60% or greater. The price and income elasticity estimates are valuable because they allow a grantor, for example, to estimate the likely reaction of grantees to several different types of grants. Further, the ability to disaggregate the factors influencing the tax price variables allows for prediction of the response to different features of aid programs. When prediction is the main objective, the median voter model is a simple and accurate alternative. For these reasons, the median voter model lives on despite its shortcomings.

CRITICAL DISCUSSION

Limitations

The median voter model has important internal and external limitations. The internal limitations are contained within the logic of the model. The external limitations question the applicability of the model to research in public expenditure analysis.

Internal Limitations

The median voter model is not a normative model. The median voter, even if decisive, may not choose the socially optimal level of expenditures. The median voter theorem is plagued by limitations as a theory of democracy; recall that the restrictions for consistent outcomes include unidimensional decisions, single-peaked preferences, and the absence of agenda setting, many of which are violated in practice.

As a theory for the study of budgeting, the median voter model may not be as limited as it is for a general theory of democracy. Many economists maintain that preferences are more likely to be single-peaked regarding budgeting decisions, that is, the farther one moves from his or her most preferred position, the lower the utility ranking will be. In terms of expenditures, if more is preferred to less, then more will likely be preferred to much less. This notion of transitivity is consistent with the familiar downward sloping demand curves associated with most goods over most income ranges and consumption levels. One possible exception would be reversion budgeting, or any all-or-nothing strategies which interfere with voter transitivity.

While single-peaked preferences seem most likely in budgeting, one can easily imagine exceptions. Proponents of high spending for public education might prefer as a second choice to send their children to private schools (and thus have minimal expenditures) over moderate spending. Similarly, proponents of high spending on public safety may prefer low expenditures over moderate expenditures because they may choose to purchase home security systems, smoke alarms and other private goods as a substitute for the public good. This pattern of preferences, combined with majority voting can lead to the problem of cycling.

Government expenditures appear to satisfy the median voter condition of a continuous, quantifiable single dimension. Are budgeting decisions truly unidimensional? Spending decisions embody more or fewer dollars, but those dollars can be associated with a number of programs and activities. Recently, more attention has been paid to the shares of the budget for particular purposes (Kamlet & Mowery, 1987), which also violates the single dimension condition.

The absence of strategic behavior precludes agenda setting, sophisticated voting, vote-trading, and logrolling for consistent majority rule outcomes. Some analysts maintain that such behavior is more likely to be a problem in representative democracies than in direct, participatory spending referenda. While voting coalitions and collusion might be difficult to engineer in a direct democracy, agenda setting seems the most ubiquitous problem. Elite actors at the sub-national level (say, school district or public safety jurisdiction) can and often do control the range and number of alternatives to be considered by the voting population. The empirical literature suggests that the presence of this behavior may move outcomes away from that of the median voter.

Limitations of the model are not solely in the restrictive assumptions. One troublesome limitation has never been resolved; Romer and Rosenthal (1979) call it the "multiple fallacy." Even when supported empirically, the median voter could actually prefer twice as much of the public good, or some other multiple, without changing any of the other coefficient estimates. If the multiple fallacy occurs, studies which seem to support the median voter model could be in error; the fallacy cannot be uncovered with more sophisticated analysis.

A second limitation which persists is the use of the individual with the median *income* as indicative of the median *voter*. This step, while a convenient simplification for purposes of data collection, is not without problems as a means of operationalization. First of all, it presumes full voting participation in the democratic process, which is far from accurate. Exacerbating this problem are the observations that participation rates tend to be highest in national elections (where choices are most "bundled" and preferences are likely to have multiple peaks), and lowest in local elections (where the median voter model has the most explanatory power and the greatest empirical support). Further, using the median income as a proxy for the median voter presumes that the income distribution mirrors policy preferences. Income is one way in which preferences may be sorted, and it varies predictably with other preference indicators: age, housing, poverty status, education, etc. While the distributions of income and policy preferences might have significant overlap on many issues, they are certainly not identical. Clearly, the median voter model is limited by the fact that the median voter might not have the median income in a particular jurisdiction. Bergstrom and Goodman (1973) confront this limitation explicitly in their study, delineating the sufficient conditions for the median voter to be the individual with the median income. These conditions are fairly restrictive, but not unreasonable.

A third limitation is the assumption that socio-economic variables such as the percentage of the population that is poor, elder, young, etc., are relevant. Strictly speaking they do not affect the preferences of the median voter. They are instead proxies for the preferences of the median voter and of the community. Thus, in practice, the logic of the median voter model is further eroded. Holtz-Eakin (1992) suggests a model that may explain the persistent significance of these variables. In that model, elected officials make budgetary decisions with a goal of maximizing votes. Voters compare the difference in the budgetary decisions of current and potential officials and their most preferred bundle of goods and vote for the candidate who is closest to their preferred bundle. To the extent that socio-economic variables characterize these preferences, these variables are expected to be significant. The Holtz-Eakin model is one potential improvement on the median voter model. Certainly any such model would need to address this issue.

External Limitations

The literature discussed here suggests that this model has the best empirical fit where the government analyzed is a direct democracy, or a single service district. As these characteristics are closest to the assumptions of the model, this is in one sense comforting. However in another sense, these limitations are troubling. Numerous studies have used such a model for representative, multiple service districts, such as cities, counties and states. Where they make an explicit comparison to another model, the results do not give strong support for the median voter hypothesis (Mueller, 1989, p. 193).

Given that we know that there are other factors – such as the structure of institutions, the process of consideration, and the distribution of political power to name a few – that can at least explain budgetary outcomes, it seems short-sighted to exclude them. Even the model's proponents recognize the need to include other variables that capture the effects of the public budgetary process. As Inman (1979) writes,

> It is apparent that there must be something more to local fiscal choice than static individual utility maximization ... [T]he hypothesized process of political choice, reduced to an 'as if' maximization of an *individual* preference function, must strike political scientists and politicians as a major departure from reality (Inman, pp. 293–294, emphasis is original).

The median voter model abstracts from all politics. The concern is that taking the next step to incorporate features of the budgetary process that we know help explain budgetary outcomes cannot be done without significantly altering and complicating the model. Such variables tend to be added in an *ad hoc* fashion, deviating from the internal logic of the model, with the unsatisfying justification that they have previous empirical support and enhance the explanatory power of the model. For example, Ladd and Yinger (1989) begin with a demand function, then incorporate political institutions with a series of variables (a lagged adjustment factor, demographic variables, and dummy variables for form of government and presence of a property tax limitation). While the result of their approach is a useful model, it is not internally consistent. The median voter model is a model of *individual* demand, therefore including political influences simply does not fit within the model. To attempt to articulate a hybrid or grand model subsuming both the median voter model and a political model does not seem possible if coherence remains a requirement of a good model.

Ingerberman and Inman (1988) argue that economic models like the median voter model have not done justice to the importance of conflict and cooperation in public budgeting. They write, "[w]e need to understand political institutions and to know fiscal institutions, for it is institutions and history, along with tastes and technology, which define the stable [political] equilibrium." (p. 143). Break (1980) puts it even more simply: "Utility maximization may well be the main motivating force behind private decisions about consumption, but it need not be so for choices of government expenditures where considerations of the general public good may weigh importantly with many voters and other participants." (p. 90, footnote 31).

One logical extension is suggested by Rubinfeld (1987). He argues for modeling the demand of groups of voters in a community, rather than a single median voter. This is the approach of Gramlich and Rubinfeld (1982). While this requires different and probably more data, it not only addresses likely econometric problems of aggregation bias in the median voter model, but more importantly it begins to transform the model into one that incorporates politics in a more fundamental way. But while it more explicitly recognizes differences among the groups, it assumes homogeneity within any group and also does not incorporate the possibility that the influence of the group is not directly related to its size, as Baumgardner (1993) argues. With further adaptation though, a political economy model such as this seems promising.

Because the median voter model is a model of individual demand for public services, a way to incorporate other influences without running contrary to the internal logic of the model would be to include a model of the supply of public services. The resulting budgetary outcome is determined by the supply-demand interaction. Thus the median voter hypothesis is abandoned, although not the median voter model. Bailey and Connolly (1998) comment that, "[w]hilst there has already been substantial theoretical development on the supply side, further progress could be made and there is huge potential in the development of a combined demand-side/supply-side model" (p. 357). One model of the supply of public services that has been commonly used by public choice theorists and others is Niskanen's (1971, 1975) "greedy bureaucrat" model. As this model is discussed in detail elsewhere in this volume, it will not be detailed here. While logically this marriage of these two models might make sense, we believe that Niskanen's model has not stood up to empirical testing (Blais & Dion, 1991; Bartle & Korosec, 1996), and further that it presents only a very limited range of bureau typologies. A second group of models of bureau supply have been presented (Romer & Rosenthal, 1978; Dunleavy, 1991; Holtz-Eakin, 1992), which offer some attractive features, although these models are often complex and difficult to operationalize. While they would likely improve the model in specific situations, they are less general. Still, some version of one of these approaches might be a more appropriate model in this tradition.

A third model of public goods supply is given by the Tiebout (1956) model. If individuals move to the locality that provides their most preferred bundle of public goods and tax levels, then the supply and demand of public goods is observable and measurable, because citizen preferences are met by the governments. However as Goldstein and Pauley (1981) have pointed out, this suggests that there is no need to estimate demand for public goods, as the levels provided are those demanded by the mobile citizens. They show that if this phenomenon is at work, one would expect bias in the estimates of a median voter model. Therefore, while the Tiebout model does include the supply side, it is a competitor to this model rather than a complement.

A fourth compelling model of public goods supply is that presented by Baum (1986) and Duncombe (1991). An explicit cost function is estimated using variables such as service outcomes, factor prices, population, and environmental variables that affect the cost of providing the service(s). Then the simultaneous supply-demand determination of output is modeled. This model is empirically-based, imposing few assumptions and restrictions. It is implicitly assumed that efficient production technology is used. This approach not only is a more complete economic model, it also allows for estimation of returns to scale, factor substitution, and the relation of inputs to costs and public service outcomes. Further, Duncombe (1991) finds that the more comprehensive supply-demand model he estimated suggests that the "flypaper effect" was due to the mis-specification inherent in a demand-only model. It is possible that the results of single equation models yield biased estimates of other coefficients as well.

FUTURE RESEARCH POTENTIAL

Despite its replete limitations, the median voter perspective does have some epistemological merit for budgetary researchers. If the median voter model can explain the demand side of public goods output, then, as Holcombe (1989) suggests, this model can be paired with a model of the supply of public goods to create a comprehensive model that links public opinion and budgetary outcomes.

But then is this a satisfactory approach? Both evidence and intuition suggest that something beyond a demand-supply framework is needed. Specifically, the effects of institutions, interest groups, history, and the process of consideration need to be integrated into the model, not in an after-the-fact way, but as fundamental tenants of the model. Bailey and Connolly (1998) write, "[u]ltimately, however, the multidisciplinary context of local government must be recognized . . . resort having to be made to institutional economics if a more comprehensive explanation of local government expenditures is ever to be provided by economic theory" (pp. 357–358). Certainly this approach is of interest to scholars of public administration and political science. While economists have striven mightily to incorporate the influence of institutions and processes into the median voter model, this is not their comparative advantage and as a result, our basis of knowledge of public spending is incomplete. It seems that it is now the time for public administration scholars to develop empirically testable models that borrow the insights of the median voter model but ground them in an alternative model based of institutional influence on budgetary outcomes.

What role will the median voter model play in future budgeting research? Because of its suitability for empirical testing, the median voter model may persist as the framework for statistical analysis of budgetary outcomes, expenditure analysis, debt financing, and fiscal federalism, especially when applied to single-purpose local governments. Where prediction and generalization is more important than explanation and context it may continue to be used. However, its limitations suggest that its most common potential is that of providing a backdrop for testing other models. Augmented median voter models typically add one or more institutional or contextual variables that are not associated with the median voter model. Statistical or substantive significance of these variables allows for identifying the effect of these variables on budgets.

While it would be a loss to discard a model that, using simple assumptions, yields a high explanatory power, explanatory power is only one of the desirable qualities of a model. In abstracting from the context of budgetary politics, the median voter model eschews explanation and description. These are important qualities of a model. As Inman says, we need to know why a theory works. The admirable simplicity of this model is not enough to overcome the omission of a realistic model of institutions. If this can be done in concert with this model, it would be an important advancement. If not, it may spell the eventual doom of the median voter model in public budgeting research.

ACKNOWLEDGMENT

Paula S. Kearns wishes to thank her husband Brian and son Ethan for their patience and support with her many professional obligations.

NOTES

1. Pareto-efficiency refers to a point from which no one can be made better off without making someone else worse off. A decision is said to be Pareto-improving if at least one person can be made better off without making anyone else worse off. If all individuals are sincere and envy-free, then unanimity assures that the Pareto condition is satisfied.

2. Atkinson and Stiglitz (1980, p. 306) generalize the concept of peak to multi-dimensional decisions. Single-peakedness is the assumption that assures a stable majoritarian outcome if all alternatives are considered.

3. Both authors of this chapter can plead guilty to this approach.

4. This equation, following Bergstrom and Goodman (1973), is based on demand for a publicly provided good: $Go = G/N^{\cdot g}$, where G is the observed level of the public good, N is the population of the jurisdiction, and g characterizes the "publicness" of the good G on an interval from 0 to 1. When $g = 0$, G is a pure public good; when $g = 1$, G is a private good. Since empirical work estimates G, not Go, the substitution $\ln G_O = \ln G - g \ln N$ can be made. Similarly, the price of one unit of the good, p^0_G, is N^g times the tax price to each individual, P_G. So, $P^0_G = P_G N^g$ allows the following substitution:

$\ln P^0_G = \ln P_G + g \ln N$. This restricts the coefficients to: $B_4 = g(1 + B_3)$ in Eq. (1) in the text.

REFERENCES

Atkinson, A. B., & Stiglitz, J. (1980). *Lectures on public economics*. New York: McGraw-Hill.

Bailey, S. J., & Connolly, S. (1998). The flypaper effect: Identifying areas for future research. *Public Choice, 95*, 335–361.

Barr, J. L., & Davis, O. A. (1966). An elementary political and economic theory of local governments. *Southern Economic Journal, 33*, 149–165.

Bartle, J. R., & Korosec, R. (1996). Are city managers greedy bureaucrats? *Public Administration Quarterly, 20*, 89–102.

Baum, D. (1986). A simultaneous equations model of the demand for and production of local public services: The case of education. *Public Finance Quarterly, 14*, 157–178.

Baumgardner, J. R. (1993). Tests of median voter and political support maximization models: The case of federal/state welfare programs. *Public Finance Quarterly, 21*, 48–83.

Bergstrom, T. C., & Goodman, R. P. (1973). Private demand for public goods. *American Economic Review, 63*, 280–296.

Blais, A., & Dion, S. (1991). *The budget maximizing bureaucrat: Appraisals and evidence*. Pittsburgh, PA: University of Pittsburgh Press.

Borcherding, T. E., & Deacon R. T. (1972). The demand for the services of non-federal governments. *American Economic Review, 62*, 891–906.

Bowen, H. R. (1943). The interpretation of voting in the allocation of economic resources. *Quarterly Journal of Economics, 58*, 27–48.

Break, G. F. (1980). *Financing government in a federal system*. Washington D.C.: The Brookings Institution.

Craig, S. G., & Inman, R. P. (1986). Education, welfare and the "new" federalism: State budgeting in a federalist public economy. In: H. S. Rosen (Ed.), *Studies in State and Local Public Finance* (pp. 187–227). Chicago: The University of Chicago Press.

Duncombe, W. (1991). Demand for local public services revisited: The case of fire protection. *Public Finance Quarterly, 19*, 412–436.

Duncombe, W. (1996). Public expenditure research: What have we learned? *Public Budgeting & Finance, 16*, 26–58.

Dunleavy, P. (1991). *Democracy, Bureaucracy and Public Choice*. New York: Prentice Hall.

Goldstein, G. S., & Pauly, M. V. (1981). Tiebout bias on the demand for local public goods. *Journal of Public Economics*, *16*, 131–143.

Gramlich, E. M., & Rubinfeld, D. L. (1982). Micro estimates of public spending demand functions and tests of the Tiebout and median-voter hypotheses. *Journal of Political Economy*, *90*, 536–560.

Grosskopf, S., & Hayes, K. (1983). Do local governments maximize anything? *Public Finance Quarterly*, *11*, 202–216.

Henderson, J. M. (1968). Local government expenditures: A social welfare analysis. *Review of Economics and Statistics*, *50*, 156–163.

Hines, J. R., & Thaler, R. H. (1995). Anomalies: The flypaper effect. *Journal of Economic Perspectives*, *9*, 217–226.

Holcombe, R. G. (1989). The median voter model in public choice theory. *Public Choice*, *61*, 115–125.

Holtz-Eakin, D. (1992). Elections and aggregation: Interpreting econometric analyses of local governments. *Public Choice*, *74*, 17–42.

Ingerberman, D. E., & Inman, R. P. (1988). The political economy of fiscal policy. In: P. G. Hare (Ed.), *Surveys in Public Sector Economics* (pp. 187–222). New York: Oxford Press.

Inman, R. P. (1978). Testing political economy's "as if" proposition: Is the median voter really decisive? *Public Choice*, *33*, 45–65.

Inman, R. P. (1979). The fiscal performance of local governments: An interpretive review. In: P. Mieskowski & M. Strasheim (Eds), *Current Issues in Urban Economics* (pp. 270–321). Baltimore: Johns Hopkins University Press.

Kamlet, M. S., & Mowrey, D. C. (1987). Influences on executive and Congressional budgetary priorities, 1955–1981. *American Political Science Review*, *81*, 155–178.

Ladd, H. F., & Yinger, J. (1989). *America's ailing cities: Fiscal health and the design of urban policy*. Baltimore: The Johns Hopkins University Press.

Niskanen, W. A. (1975). *Bureaucrats and politicians. Journal of Law and Economics*, *18*, 617–643.

Niskanen, W. A. (1971). *Bureaucracy and representative government*. Chicago: Aldine-Atherton Press.

Mueller, D. C. (1989). *Public choice II*. New York: Cambridge University Press.

Pommerehne, W. W. (1978). Institutional approaches to public expenditures: Empirical evidence from Swiss municipalities. *Journal of Public Economics*, *9*, 163–201.

Pommerehne, W. W., & Frey, B. S. (1976). Two approaches to estimating public expenditures. *Public Finance Quarterly*, *4*, 395–407.

Romer, T., & Rosenthal, H. (1978). Political resource allocation, controlled agendas, and the status quo. *Public Choice*, *12*, 27–43.

Romer, T., & Rosenthal, H. (1979). The elusive median voter. *Journal of Public Economics*, *12*, 143–170.

Rubinfeld, D. L. (1977). Voting in a local school election: A micro analysis. *Review of Economics and Statistics*, *59*, 30–42.

Rubinfeld, D. L. (1987). The economics of the local public sector. In: A. J. Auerbach & M. Feldstein (Eds), *Handbook of Public Economics* (Vol. 2, pp. 571–645). Amsterdam: North-Holland.

Thompson, F. (1998). Public economics and public administration. In: J. Rabin, W. B. Hildreth & G. J. Miller (Eds), *Handbook of Public Administration* (2nd ed., pp. 995–1063). New York: Marcel Dekker.

Tiebout, C. M. (1956). The pure theory of local expenditures. *Journal of Political Economy*, *64*, 416–424.

Turnbull, G. K., & Mitias, P. M. (1999). The median voter model across levels of government. *Public Choice*, *9*, 119–138.

6. PUBLIC CHOICE THEORY AND PUBLIC BUDGETING: IMPLICATIONS FOR THE GREEDY BUREAUCRAT

John P. Forrester

ABSTRACT

Public choice theory (PCT) for years has been relied on by conservative economists to describe bureaucrats as rational self-maximizers. This chapter examines the relevance of PCT to the study of public budgeting. Here I suggest that while there is a rift between PCT and the approach used by public budgeters to study budgetary behavior, there is much that the students of public budgeting can learn from public choice theory.

INTRODUCTION

Throughout its history, the field of public budgeting has gone through periods of self-reflection to assess the directions for future research. Today, budgeting has entered such an era. Most recently, the questions have centered on the cascading levels of responsibilities of budget offices and budgets (see Berman, 1979; Pearson, 1980) – which have become entrenched in addressing the important but perhaps peripheral issues of budgeting – and the parallel bogging down

Evolving Theories of Public Budgeting, Volume 6, pages 101–124.

ISBN: 0-7623-0790-0

of budget theory, where the whole notion of budgeting is lost in a quagmire of politics and bureaucracy. Consequently, some scholars have argued that budgeters need to return to addressing fundamental questions of theory (Schick, 1988; Kiel & Elliott, 1992; Forrester & Adams, 1997).

The purpose of this chapter is to continue in the footsteps of this most recent trend by asking how the study of public budgeting and bureaucrats who make budgetary decisions can be advanced from yet another provocative perspective, that of public choice theory (PCT). Why PCT? Regardless of one's political or policy preferences, at the state and local level the theoretical grounding for the tax and expenditure limitation movement of the last two decades and the debates over school choice can be traced to Niskanen and other proponents of public choice. At the federal level, the cuts proposed by the Reagan administrations and even the cuts in growth proposed by the Clinton administrations can be traced back to arguments of public choice. Simply put, public choice is affecting government resource allocation.

Like public budgeters, public choice theorists have long addressed the issue of resource allocation in government and factors affecting that allocation, but they have not always come at the issue from the angle – historically, conceptually, or methodologically – taken by budgeters. These differences, however, are moderated by critical commonalities shared by public budgeting and public choice. By examining the assumptions and the evolution of public choice theory, this study hopes to provide the reader with a deeper appreciation of public budgeting theory and practice, in particular its implications for the bureaucracy.

The discussion will proceed by first defining public choice and presenting its fundamental assumptions. Second, the great divide that has traditionally separated public choice and public administration generally and public budgeting specifically will be reviewed. In spite of the depth of the divide, there are significant issues on which the fields are in relative harmony. Because this is what we hope to learn from, these parallels will be explored next. With the conceptual foundation laid down we can begin to hypothesize about the implications that taking a public choice perspective has on bureaucracy. Lastly, some concluding remarks and research questions are presented.

PUBLIC CHOICE: WHAT IS IT?

What is public choice? Public choice is the "analysis of choices made by public institutions" (Borooah, 1996, 43), or "the economic study of non-market decision making, or simply the application of economics to political science" (Mueller, 1989, 1). In this vein, public choice theorists assume individuals behave in a rational and utilitarian manner, reveal their preferences in a market-like fashion (e.g. exchange of votes, Tiebout-like voting), and ask traditional pricing-theory questions (e.g. Are decisions Pareto efficient?) (Mueller, 1989).

The public choice approach is within the 19th century neoclassical economics tradition (because of its emphasis on the individual), but is more deeply rooted in the political economy of old, which included both the study of government behavior and the classical economics of Smith, Mill and Marx (van Winden, 1988). During the early part of the 20th century, as Mueller clearly points out, public choice as a field began with a disenchantment towards market socialism (see Bergson, 1938; Arrow, 1951; Samuelson, 1947). Today the field of public choice has come to address a broad range of topics, including voting rules, allocational efficiency versus redistribution, majority rule versus other forms of rule, theories of clubs and public goods, legislative decision making, rent seeking controls through regulation and other government actions, bureaucratic behavior, political business cycles, government growth, and interest group actions. The literature examining these issues reflects attempts to develop theories, positive and normative, of resource allocation. Given the focus of this chapter, we will restrict ourselves only to discussions of bureaucratic behavior and the influence it may have on the allocation of public resources and public budgeting more generally.

One of the most vocal normative theorists, Vincent Ostrom, has argued that public choice reflects the principles of self-governance and self-interest as charted in the *Federalist Papers* (see Ostrom, 1974; Lovrich & Neiman, 1984) and as characterized by Tocqueville (1835 and 1840: I, 64 as cited in Ostrom, 1974; see Wicksell, 1896 as cited in Mueller, 1989). If the principles are re-iterated among the units of government, then, Ostrom suggests, "public administration is confined to circumstances where centralization and hierarchy can be held to a minimum" (Ostrom, 1974, 92). In this system, authority is fragmented within and among governments and overlaps jurisdictions to guard against the centralization of power (Ostrom, 1974). Decisions made by governments of this system promote a better society by ensuring and protecting the liberty of individuals. Where the bureau is the monopolist supplier, restricts information regarding its true costs, and "is institutionally allowed to make take-it-or-leave-it budget proposals," bureaucracies may have the power to obtain excessively large budgets (Mueller, 1989, 255; Also see Niskanen, 1971 and 1975). It is this normative, paradigmatic perspective of Ostrom that has perhaps set public administrators most firmly against the tenants of public choice.

Ostrom's normative perspective of public choice views bureaucracy as evolving into a Weberian-based polycentric democratic theory of administration, the public choice alternative to what Ostrom calls the Wilsonian-based monocentric bureaucratic theory of administration (Ostrom, 1974).[1] The democratic theory of administration is characterized by:

(1) Individuals who exercise the prerogatives of government are no more nor no less corruptible than their fellow men.
(2) The exercise of political authority – a necessary power to do good – will be usurped by those who perceive an opportunity to exploit such powers to their own advantage and to the detriment of others unless authority is divided and different authorities are so organized as to limit and control another.
(3) The structure of a constitution allocates decision-making capabilities among a community of persons; and a democratic constitution defines the authority inherent in both the prerogatives of persons and in the prerogative of different governmental offices so that the capabilities of each are limited by the capabilities of others . . .
(4) The provision of public goods and services depends upon decisions taken by diverse sets of decision makers and the political feasibility of each collective enterprise depends upon a favorable course of decisions in all essential decision structures over time. Public administration lies within the domain of politics.
(5) A variety of different organizational arrangements can be used to provide different public goods and services . . .
(6) Perfection in the hierarchical ordering of a professionally trained public service accountable to a *single* center of power will reduce the capability of a large administrative system to respond to diverse preferences among citizens for many different public goods and services and cope with diverse environmental conditions.
(7) Perfection in hierarchical organization accountable to a *single* center of power will *not* maximize efficiency as measured by least-cost expended in time, effort, and resources.
(8) Fragmentation of authority among diverse decision centers with multiple veto capabilities within any one jurisdiction and the development of multiple overlapping jurisdictions of widely different scales are necessary conditions for maintaining a stable political order which can advance human welfare under rapidly changing conditions (Ostrom, 1974, pp. 111–112).

This public choice perspective on the administration of government was offered to challenge the apparently prevailing values espoused by Woodrow Wilson of an administration that was separate from politics, based on universal principles of good administration, and had a misconception on efficiency (Ostrom & Ostrom, 1971). Accordingly, such a method of administration would be organized in a hierarchically and non-competitive manner, "with different levels of responsibility and different scopes of activity" (Stevens, 1993, 270). Such a structure, however, would not only be inefficient, but it would "be an invitation to exploitative behavior" (Niskanen in Stevens, 1993, 270). The polycentric approach would tend to inefficiencies and exploits by presuming that individuals operate in their self-interest, and only structures that accept that fact (structuring to facilitate competition) can operate efficiently. Decentralized, flat and competitive organizations were the order of the day (see Downs, 1967).

PUBLIC CHOICE AND ITS CONFLICTS WITH PUBLIC BUDGETING

Some Traditional Assumptions of Public Choice

The force behind public choice theory traditionally is the perceived validity of the principles that seem to guide the economic marketplace. These principles, or assumptions, include self-interest, exchange, methodological individualism, Pareto optimality, and a preference for methods and techniques of economic analysis (DeGregori, 1974; van Winden, 1988; Udehn, 1996).

Self-interest is the "main ingredient in the theory of public choice" (Udehn, 1996, 60; also see Niskanen, 1971; Buchanan, 1986, 1989; Buchanan & Tullock 1965). Under this assumption, the individual is the unit of analysis and "[t]raditional 'economic man' is replaced by 'man: the decision maker'" (Ostrom & Ostrom, 1971, 205). People, whether individually or collectively, are thought to act in ways that maximize their individual welfare or utility (see Olson, 1965), even though there is no scientific principle to follow in determining the worthiness of such actions (Mikesell, 1991). As Stevens argues:

> To an economist, thinking about *self-interest* starts with the idea of individual utility. Economists say that 'people have utility functions,' meaning that people assess and define their own needs and preferences. This doesn't mean that they are necessarily selfish or benevolent. A person who donates all his wealth to charity and has few regrets, for example, might be maximizing his self-interest. So too might the miser who never shares with others. Economists would regard both individuals as rational if their actions are consistent with their goals and objectives (p. 96).

For government, this means that bureaucrats will strive to maximize their budgets (see Niskanen, 1971) or their slack – the difference between the bureaucrat's revenue and the minimum cost of production (Wyckoff, 1990,[2] see also Cyert & March's (1963) discussion of organizational slack; Migué & Bélanger's (1974) intriguing discussion of discretionary profit (with a following commentary by Niskanen, 1974); and Brenton & Wintrobe, 1975) – for reasons of self-interest, including salary needs, power, prestige, patronage and reputation (Niskanen in Stevens, 1993).

The principle of exchange ties public choice most directly to theories of the marketplace. Public choice theorists see the political marketplace based on a science of exchange of public goods by self-interested bureaucrats, along with politicians, interest groups and voters (see Udehn, 1996). This exchange, however, unlike that in the private sector, tends to depend upon only one, monopolistic seller – the government (see Buchanan & Tullock, 1965); and man is driven by economic rather than political values (Buchanan, 1986). From the exchange, preferences of the actors are aggregated through the political process into the publics' choice. As Udehn (1996, 116) points out, the market metaphor for political theory was produced by public choice theorists (see Elster, 1986) and continues today as a fundamental core assumption of public choice (Buchanan, 1988).

The third fundamental assumption driving the theory of public choice is that analysis should take the perspective of the representative individual (Buchanan, 1966; Ostrom & Ostrom, 1971). Here, the ends of the state are the ends of the individuals within that state, where, "State decisions are, in the final analysis, the collective decisions of individuals" (Buchanan, 1949, 498). This perspective, "methodological individualism," may even be considered the starting point of public choice theory (van Winden, 1988, 11).

Public choice theorists also believe in both Pareto optimal solutions and the power of the competitive market place to optimally allocate resources (i.e. Coase's Theorem, see DeGregori, 1974). According to the Pareto criterion, "if at least one person is better off from a policy action and no person is worse off, then the community as a whole is unambiguously better off for the policy" (Mikesell, 1991, 6; Pareto, 1927, 617–618). Pareto optimal solutions are preferred to solutions based on paternalism (where individuals are not seen to be the best judges of their own welfare) or "conceptions of social welfare which transcend individuals" (Heap et al., 1991, 342–345).

Lastly, public choice theorists prefer to use methods and techniques of economic analysis in their study of governmental decisions (Van Winden, 1988). Economic-type graphs and market-based discussions are found throughout early works by Tullock (1970, 1967) and even in his more contemporary book on economic hierarchies (Tullock, 1992), where more mathematically-based presentations are found in two of his co-edited books (Brady & Tullock, 1996; Tullock & Wagner, 1978). For a broad overview of the methods used by public choice theorists and analysts, just take a cursory view of several issues of the journal *Public Choice*.

The Conflicts

The assumptions that public choice theorists make when proceeding with their analyses are clearly based on historical principles of private sector economics. Their usefulness as principles of public sector fiscal and budgetary behavior, as with most theories of budgeting and resource allocation is not clear. Some testing of the principles has been done, but the results are not conclusive. In other areas such as the hypothesis that bureaucratic power leads to larger budget/slack, tests have yet to be conducted (Mueller, 1989). Even without these tests, many have come to question whether such principles are even applicable to the public sector (Golembiewski, 1977a, b; Stillman, 1976, 1978; Waldo, 1952). Here, we identify five categories of differences between public choice as traditionally characterized and public budgeting: appropriateness of their economics framework, analytical methodologies, bureaucratic assumptions, value orientations, and congruence with reality. In each case, because of the tentative and incomplete findings of the literature, the differences should be taken as a matter of degree rather than absolute.

Economics Framework

While the study of public budgeting is fiscally oriented, the economics framework used by public choice theorists may have limited value in the overall study of budgeting. First, the exchange metaphor may be generally misleading (Udehn, 1996) and descriptively limited (Balbus, 1975). Second, the approach assumes that individuals are the best judges of their own needs. Given their limited cognitive abilities (see Simon, 1945; Lindblom, 1959), their assessments are more likely to be limited and marginal rather than rational and comprehensive. Third, in order to maximize one's utility there must be a consistent and logical ordering of objectives, and an individual's actions must be consistent with those objectives. Unfortunately, self interest and goals are not always clear, and objectives in government are rarely ordered, let alone rigorously defined (see March & Olsen, 1976; March, 1978). In turn, determining congruence between objective and action is likely in relatively few instances (congruence is most likely to occur in enterprise functions and investment decisions). Fourth, the public choice approach dwells so much on the economics of government allocation decisions, that other important disciplinary perspectives are often forgotten. Political, sociological, and management perspectives, for example, have been used in the budgeting literature to examine such timely issues as retrenchment (see Levine et al., 1981; Levine, 1979; Rubin, 1985), backdoor appropriations, and rebudgeting (Caiden & Wildavsky, 1974; Caiden, 1980; Lauth, 1988; Forrester & Mullins, 1992; Forrester, 1993). Finally, public choice theory has "neglected the negative externality of the personal freedom of the powerful to deny basic economic goods and thereby curb the freedom of the less privileged" (DeGregori, 1974, 219).

Analytical Methodologies

The methods of analysis that public choice theorists use also are generally very different from those used by public budgeters. This is true at both the conceptual and operational or modeling levels. Conceptually, public choice theorists model behavior from a very conservative constitutional perspective (Lovrich & Neiman, 1984, Ostrom, 1974, Ch. 4 and 5) rather than from a bureaucratic politics or management perspective that budgeters would use. Operationally, public choice theorists tend to propose and test very mathematically rigorous models based on economic assumptions whereas research on public budgeting often makes use of case studies, surveys, and relatively basic descriptive statistics. Less frequently found are analyses grounded in economic theory, and where such analyses are found they tend to address the revenue side of budgeting rather than the expenditure side. The reason is that public budgeting research tends to incorporate theories and perspectives from the various fields, including sociology, psychology, organization dynamics, systems theory, and political science, not just economics.

Bureaucratic Assumptions

Public choice and public budgeting also tend to make fundamentally different assumptions about the bureaucracy, or more accurately, public choice makes very clear assumptions about the bureaucracy whereas the assumptions made by public budgeters are rarely pointed out. Niskanen (1971) and Olson (1965), for instance, argue that individuals join government agencies or other volunteer or nonprofit organizations to maximize their own self-interest. To further their self-interest, in the context of the organization, the bureaucrats will try to control the use and dissemination of information (resulting in an asymmetry of information vis-à-vis the legislature) and inefficiently maximize their budgetary resources (Hyman, 1990, Lovrich & Neiman, 1984, Niskanen, 1971, 1975). Their models of bureaucratic behavior, then, assume that conflict (Niskanen's 1971 model; Stevens, 1993, also see Crozier's 1964 discussion of uncertainty, power and conflict), not cooperation ("iron triangle" and Helco's 1977, 1978 "issue networks") portrays the relationships between bureaucracies and others. For instance, Stevens (1993, 269) comments on Niskanen's model: "It was the first major deductive model of bureaucracy in representative government, and it was based explicitly on conflict between legislative and administrative branches of government." It has been long-lasting because of its focus on "three crucial elements of the relationship between elected and appointed suppliers. These elements are bilateral monopoly, asymmetric information, and budget maximization."

Public administrators, however, have tended to take a different perspective. Both cooperation and conflict are ever present, and the relative dominance of one or the other is determined by several factors – political, economic and bureaucratic. Axelrod argues, for instance, that in spite of "the new superstructures of congressional budgeting [of the early 1990s], the old legislative workways persisted. Alliances of congressional committees, interest groups and administrators developed many of the political agendas." The iron triangles, he continues, have "left their imprint on policies and budgets even under the axe of deficit control" (1995, 205). But the spirit of conflict between all parties of the iron triangle and the chief executive is also variously portrayed by Rubin's examination of the Reagan Administration's efforts to cut the Bureau of Health and Planning, the Employment and Training Administration, the Community Development and Planning Program, and the Urban Mass Transportation Administration (Rubin, 1985). In short, the tides of cooperation and conflict and resource shortages and unexpected growths suggest that bureaucracies are not predestined to expand and become engorged. Goodsell, in his 1983 book *The Case for Bureaucracy*, contends that evidence shows that bureaucracies are not always expanding, in fact sometimes they even die (see Kaufman, 1976).

Because of their assumptions about the bureaucracy, public choice theorists (like accountants) tend to look at the bureaucratic-legislative world in the context of principals and agents, where the main issue is who controls whom. In Niskanen's world the control of information by the bureaucrats and the monopolistic powers held by the bureaucracy in service delivery (Hyman, 1990) suggests that bureaucrats, not legislators, hold the upper hand. In the public budgeter's world, to the degree the iron triangle concept is accepted as the norm, dominance, while discussed, is rarely modeled explicitly. The remaining options are for the legislative body to dominate budgetary and programmatic negotiations, where "the legislative structure, particularly the committee system and majority rule, facilitates the pursuit of self-interest by legislators and leads to outcomes that are stable and predictable" (Stevens, 1993, 290), and for the interest groups to dominate the negotiations, which may occur in part in states that have imposed legislative term limits.

Value Orientations

Supporters of public choice also tend to have very different value orientations than public administrators and public budgeters. First, traditional public choice theorists put much currency in the concepts of expected utility and economic man. Most recently, Rowley, Schneider and Tollison, co-editors of the twenty-fifth anniversary issue of the journal *Public Choice*, argued that there is no reason for public choice theory to abandon its faith in the expected utility model or its belief that "*Homo economicus* both sets the rules and maximizes subject to the constraints imposed" (Rowley et al., 1993, 2–4; the paradigm is expressed most vividly by Ostrom, 1974; 1977). While this view may hold among such traditionalists, it has come under fire from within the ranks of public administration (e.g. Golembiewski, 1977a, b; Stillman, 1976, 1978), economics (e.g. DeGregori, 1974), sociology (e.g. Udehn, 1996),[3] and even from public choice advocates who believe that for public choice to become politically and policy relevant it must expand into psychology (Frey, 1993)[4] and social psychology (Buchanan, 1993).[5]

Reality

A fifth major difference between public budgeting and public choice is the degree to which their arguments may reflect reality. Clearly, public choice advocates generally believe that their models and empirical studies reflect the real world of political and bureaucratic decision-making (again, see Mueller's thorough review of the literature; other current examples include Ruggiero et al., 1995 – use of liner programming to estimate technical efficiency; and DeBorger et al., 1994). There are exceptions, however. Stevens (1993, 264), for instance, contends that "Only Niskanen's (1971) hypothesis that bureaus try to maximize budgets . . . has been a major step toward understanding how administrative government works; even so, his model remains largely untested." Even Frey (1993, 96) suggests that "While the success of public choice within positive economics and political science is beyond doubt, its impact on normative economics and on actual policy making leaves much to be desired.[6] When it comes to making recommendations, the standard approach in economics still follows the traditional welfare view . . ." (see also Udehn, 1996). And the debates between Ostrom and Golembiewski and between Ostrom and Stillman highlight the doubt about the applicability of the public choice paradigm to the understanding of administration in the public sector generally. Public choice has been very rigid in its principles and paradigmatic in its orientation, to the point where even leaders of the movement are very resistant to accepting criticism from within (see Rowley et al.'s arguments against the proposals put forth by Frey & Buchanan).

Research on public budgeting, contrary to that found in public choice, is not guided by a central paradigm, its principles are uncertain, and its focus is very pragmatic. Budgeters and researchers in budgeting may consider themselves either incrementalists or rationalists, but these represent neither paradigms nor clear principles of resource allocation. And, ironically, because there is no guiding paradigm and the budgeting principles that exist are few and uncertain, the validity and meaning of much budgeting research is unclear.

PUBLIC CHOICE AND ITS HARMONY WITH PUBLIC BUDGETING

The academic traditions of public budgeting and public choice have clear and identifiable differences that separate the two in theory and in practice. This is clear. Nevertheless, the significance of the differences may be tempered somewhat by the many similarities that the two share. First, both have a similar fundamental interest – analyzing the choices that governments make (see Borooah, 1996). As with the differences, several of the similarities are both historic and a few are more recent (e.g. Frey, 1993; Buchanan, 1993). Public choice advocates almost uniformly believe that the underlying value that guides the choice should be Pareto optimality and expected utility from a conservative economic perspective. Public budgeters, however, find the analysis of choice to be rooted in a variety of values, including economics (see Mikesell, 1991), management science and operations research (see several of the tools examined by Stokey & Zeckhauser, 1978; McKenna, 1980), politics (Wildavsky, 1988), and even culture (Forrester & Adams, 1997). These differences, however, mask the fact that both traditions use rationally based arguments to explain resource allocation behavior, and, similar to Simon's reflection on a theory of administrative decisions, both are "preoccupied with the rational aspects of choice" (Simon, 1976, 62). The traditions, however, apply different standards and frameworks of rationality.

Second, both areas of study have similar views about the bureaucracy, at least in the areas of the politics-administration dichotomy, definitions of success, and need of governments to function economically. For instance, most advocates from each area agree that the administration of government is not really separate from the politics of government. In public choice, the bureaucracies consistently are seen to be engaged in political gamesmanship with the legislative body, strategizing on how best to use the information they possess to influence the legislature to maximize their budgets (again, see Niskanen's model). And in public administration, except for the early years, where principles of administration were presented as separate from administration (see Gulick, 1937), public administrators have argued for years that the two are not dichotomous (see Waldo, 1948; Appleby, 1949; Long, 1949; Seidman, 1980;[7] Goodsell,

1983; Redford, 1969). They have also argued quite eloquently about the games and strategies played by the executive branch (between agencies and the CEO and between the executive and the legislative branches) in budget preparation, adoption and implementation (see Wildavsky's 1988 discussion; McCaffery, 1988; Rubin, 1990; Meltsner, 1971; and most recently by Meyers, 1994).

Third, scholars of both budgeting and public choice believe that one of the important criteria agencies use to define their success is the bottom line, or the percentage increase in the budget this year compared to last year. From the public choice perspective, budget maximization, unfortunately to an inefficient degree, is a fundamental truth because individuals always behave as self-maximizers (see Haveman, 1976, Downs, 1967). Public budgeters also generally contend that agencies will try to expand their budget bases, but for slightly different reasons, including: natural tendencies of incremental growth (Wildavsky, 1988), at least at the most aggregate level of expenditures (see Davis et al., 1974); demands for either better delivery systems, increases in workloads or service expansion (see McCaffery, 1988); expanding the rights to classes of individuals to government resources (Straussman, 1988; Straussman & Thurmaier, 1989); and transferring responsibilities from one level of government to the other. Budgeters also believe, however, that expansion is not always the rule. For instance, agencies under fiscal stress often engage in cutback management (sometimes with minimal resistance) or at best may simply seek stability rather than cuts (see Rubin, 1985; Behn, 1980; Levine, 1978). Also, while expansion may be one objective, research suggests that it is only one of many objectives and it is not necessarily the most important for agency administrators (Duncombe & Kinney, 1987).

Both public budgeting and public choice scholars also agree, periodically, that government should adopt an economizing orientation. For those in public choice, this perspective is obvious. In public budgeting there is no economizing paradigm that guides all of budgeting research, but two streams of developments suggest its omnipresence. On the one hand, twentieth-century budgeting evolved from the traditions of accounting (see Rubin, 1993) that embraced line-item budgeting, which is still widely used today, and cost controls. This type of budget and accompanying decision-making process encourage decision makers to ask "How much does this item cost?," "Why do we have to spend that much?," and "Can the department spend less?" On the other hand, the value of economy has gained wide-spread acceptance academically, especially with the seminal work by Allen Schick in 1966. The article clearly indicates the significance of the economy orientation in the early twentieth century. Even more recent budgetary reforms that emphasize management/efficiency (e.g. reorganization of the Bureau of the Budget, performance budgeting, ZBB) and planning/effectiveness (e.g. PPBS, outcome based budgeting or the performance based budgeting of the 1990s), no matter their level of technical sophistication, have incorporated line-itemization as part of their core, giving decision makers the means for questioning the costs of inputs.

Finally, advocates of both traditions have reached out to theories about various agents, including the electorate, elected politicians, committees and subcommittees, political parties, bureaucrats, and interest groups to help explain resource allocation decisions of governments. In the public choice arena, the roles and significance of agents have been spelled out with great clarity (see van Winden's 1988 discussion of political parties; Downs' 1957 discussion on voters and politicians; and Niskanen's 1971 discussion of bureaucrats). The roles of agents are especially visible in the literature evaluating governmental consolidations and structures (see Marando & Whitley, 1972; Hawkins, 1966; and Ostrom, 1971). An accurate modeling of the supply and demand for government services, van Winden argues, should reflect the effects of external pressure groups on bureaucratic behavior (1988). Economic and political interests are also commonly seen as having a significant impact on local development policy in the related field of political economy (see Molotch, 1976; Schultze, 1985; Elkin, 1987; Logan & Molotch, 1987; Rubin & Rubin, 1987; Stone et al., 1986; Stone & Sanders, 1987; Stone, 1980, 1989; Spindler & Forrester, 1993).

Public budgeters also believe that budgetary decisions often hinge on the pressures and influences of these various groups. One only need to look at the research on California's Proposition 13, on managing cutbacks during times of fiscal stress, on strategies used by agencies to garner budgetary increases, and even the research on recent pressures to reform federal budgeting to get a feel for the significant roles agents play from the perspective of budgeters. The difference, however, is that assumptions about the roles in the public choice paradigm are explicitly stated (see Niskanen, 1971; Stevens, 1993, in Ch. 9),[8] theoretically based, and treated fairly consistently across most studies, whereas in the public budgeting literature the assumptions about the roles are rarely theoretically based, rarely explicitly stated, and therefore are addressed inconsistently from study to study. Some of the more notable exceptions to this argument are treatise on federal budgeting by Wildavsky (1988) and Feno (1966), and the political economy of legislators by Schier (1992).

IMPLICATIONS FOR THE BUREAUCRACY

So, what implications does public choice theory have for the bureaucracy, especially in the context of budgeting? The answer, unlike that provided by Ostrom, is not necessarily clear, but we may provide insight by first highlighting the limitations of public choice theory and second identifying the potential relevance of public choice theory to public budgeting. The limitations are presented in the context of the earlier discussion on Ostrom's arguments and the assumptions of public choice. What public choice may contribute to the understanding of public budgeting will be presented through a discussion of the general harmonies between the two as presented earlier, and a reflection of the impact of budget structure and cost structure on bureaucratic behavior.

Public Budgeting, PCT and the Greedy Bureaucrat: The Rift

The arguments of public choice theorists suggest that bureaucrats are 'greedy' to the extent that they engage in behavior that allows them to maximize their clearly articulated goals. The veracity of each argument, however, has been challenged by several academics in public administration. Here we address a few of these important presumptions and argue that either bureaucrats are not necessarily greedy, even if this greed is merely benevolent, or the greed of bureaucrats is effectively channeled by the budgetary and organizational conditions that frame government.

Ensuring Bureaucratic Accountability: Multiple and Divided Authorities

Public choice advocates argue that authority should be divided and disbursed to control and limit the power of bureaucrats, and to assure that a bureaucracy is accountable to the public. Accountability, however, can be achieved in several ways, and it may not be efficiently or effectively achieved through institutionally based fragmentation and competition. For instance, where service delivery costs decrease at the margin, such competition simply will not work-lower cost output is achieved by larger scale production. Instead, control and accountability are more effectively assured through pre-auditing procedures (e.g. a controller's check on requisitions and purchase orders), budget office expenditure controls (e.g. apportionments), and statutory restrictions (e.g. Congress may enact "statutory requirements in appropriations, such as defining periods of availability for funds, earmarking funds for certain projects or activities, and breaking accounts down into sub-accounts (activities)). In addition, Congress may include detailed instructions and earmarkings in the committee reports that accompany appropriation bills" (*National Performance Review*, 1993, 34–35). In such ways, greed tends to be effectively channeled and constrained.

The Greedy Bureaucrat: Defined by the Domain of Politics

True, bureaucrats are affected by politics, but this finding does not uniquely support the idea that public choice is the appropriate paradigm for public administration or public budgeting any more than it supports innumerable other paradigms. Politics is simply a component of nearly every activity, even some that are largely private by nature. Simply put, that public administration is political is an assertion not unique to public choice. Secondly, the claim is also of limited validity; public administration and public budgeting lie partly, not uniquely, within the domain of several fields and disciplines, including political science, accountancy, sociology and others, even economics. Consequently, the behavior of bureaucrats can only be reasonably modeled by drawing on the theories from such fields; constructing a model of bureaucratic behavior from theories limited to only one field, such as public choice, will severely limit its usefulness.

Managing the Greedy Bureaucrat: Efficiency Through Organization

Public choice advocates argue that a variety of organizational arrangements can be used to assure that goods and services are provided efficiently. Relying on various organizational structures (see Mintzberg, 1979, for instance) or conflicts (e.g. in a competitive market place) indeed may be a central strategy for efficiently delivering services. However, the effectiveness of such arrangements may be uncertain, especially if the service provision costs decrease at the margin. In such cases, there may be other ways of assuring efficiency, such as changing the budget structure and implementing appropriate reporting requirements. Such changes are currently being executed at the federal level (see NPR 1993; *Reinvention's Next Steps* 1996), state level (e.g. in Oregon – see *The Oregon Option*, 1994; and Missouri, Forrester, 1993), and even at the local level (see Osborne & Gaebler, 1992; also see the Governmental Accounting Standards Board, which has encouraged governments to measure their service efforts and accomplishments and incorporate them in their financial statements – GASB, 1987, 1994).

Assumptions of Self-interest, Methodological Individualism, Pareto Optimality, and Methods/Techniques of Economic Analysis

Each of these assumptions is fundamental to the validity of public choice and its relevance to public budgeting and the bureaucrat. Unfortunately, each is also somewhat unreliable. To assume that bureaucrats operate in their self-interest assumes that they have identified and ordered their personal goals and objectives, that they have good measures to determine the extent to which they have accomplished these desires, and that they have rigorously evaluated the data and reflected prescriptively on the findings. It also assumes that they have made a similar assessment of their agency's activities and have compared the findings of themselves with the findings of the agency. Limited human cognitive abilities virtually prohibit such analyses. Instead, bureaucrats are likely to make limited comparisons, often qualitatively, with limited information and within the confines of a few objectives. Within this restricted analytical framework, however, bureaucrats may operate to maximize their self interest, their budget, or their slack, as much of the budgeting literature argues.

Methodological individualism is also an assumption of limited value. One of the problems is simply determining who is the representative individual and what are his/her needs and wants. Who is the median voter? Who is the representative student? Who is the representative parent of students? Who is the representative unemployed person? We might be able to provide an answer to each question, IF each question is conditioned (e.g. controlling for race, age, gender, type of agency). If a representative individual cannot be characterized and identified, then it will be virtually impossible to effectively generalize about such an individual, limiting the value of public choice theory to explaining bureaucratic behavior.

Such constraints also affect pareto optimality; it is virtually impossible to identify any public policy or program that does not adversely affect someone's economic, social or psychological welfare. Municipal airports redistribute wealth from the have-nots to the haves. University-run restaurants take business away from the private sector. Even state government retirement programs take money away from potential privately operated investment programs. Instead of relying on pareto optimality to make allocation decisions, elected representatives and bureaucrats may find other criteria more reasonable criteria, such as Kaldor-Hicks (to maximize net benefits where distributional impacts are not too much of a concern, e.g. a regional airport) and interpersonal comparisons.[9] The decisions are also likely to be a function of the bureaucrat's and culture's risk preference (avoidance, neutral, or seeker); whether one is trying to maximize gains or minimize losses and opportunity costs; the quality and the cost of the information available; and a host of other factors.

Finally, proponents of public choice theory argue that methods and techniques of economic analysis are appropriate for studying the behavior of the public sector. In public budgeting, economic analysis is used primarily to analyze the revenue side, rarely is it used to study expenditures. However, because of the complex disciplinary characteristics (e.g. political, economic, psychological, accounting) of government, the study of bureaucratic behavior with respect to revenues and expenditures cannot be limited to this bag of tools if the results are to be useful to policy makers. Basically, the domain of methodological and analytical tools that are used for conducting an analysis should be determined by the issue being analyzed. For instance, we would expect bureaucrats trying to: (1) characterize or describe the causes and effects of past/current problems and polices – descriptive issues – to use a monitoring method; (2) characterize or describe the causes and effects of future problems and policies – predictive issues – to use a forecasting method; (3) determine the worth of a policy or program – evaluative issues – to use an evaluative method; and (4) recommend one or more policy/program alternatives – prescriptive issues – to use a normative method (see Dunn, 1994). Exactly which method or combination of methods will be used will be determined by very pragmatic factors, including the financial resources available, the skills of the bureaucrat (or who ever will do the analysis), and time limitations.

Public Budgeting & PCT: The Nexus

The rift between public budgeting and public choice is great, but it is not absolute. Public choice theory, or at least parts of public choice theory, may be very effective in helping public budgeters understand the behaviors of bureaucrats as they participate in the resource allocation process. Both fields are fundamentally interested in explaining governmental choice; both give little credence to the concept of the politics/administration dichotomy; both define budgetary success similarly; and both see agents as playing key roles in decision making. Because of these parallels, there is good reason to believe that public budgeting research could benefit from drawing on the strengths of public choice, much as it has with other fields and disciplines. For instance, research on budgeting generally and bureaucratic behavior in budgetary decision making specifically could advance theoretically if studies more rigorously and explicitly identified and tested various principal-agent relationships. Or directly incorporating PCT perspectives on budgetary maximization and slack maximization into research on budgeting could lead to more insight on why agency budgets grow and when they are likely to shrink. Perhaps in the context of such theories, budgeting research could make more fundamentally sound arguments regarding role behaviors during budget preparation, adoption, approval, and execution, if it rigorously and explicitly hypothesized and tested the asymmetry of information within bureaucracies and between actors in phases of the budget cycle.

More generally, the tenants of public choice may shed light on the impact of organization arrangements on bureaucratic behavior – or at least where organization structures advocated by public choice are implemented (multiple competing governments) we may be able to draw on other disciplines to figure out how bureaucrats will be affected; how bureaucrats should make decisions; and how bureaucrats interact with the executive, the budget office and the legislative body.

REFLECTION

Trying to predict or model the behavior of any person, type or group of persons, or markets has been a challenge for anyone willing to venture into such chilling waters. In this chapter we did little more than dip our big toe – several of the limitations and possibilities of public choice theory vis-à-vis the budgeting bureaucrat were reflected on. The arguments and assumptions of public choice may seem anathema to most scholars in public administration or at best a "hmmm, interesting" to scholars and practitioners in budgeting, but in this chapter we argue that writing-off the field, along with its assumptions and methods, is a mistake. Instead, there are conditions under which public choice is reasonably applicable to government. Today, several governments are experimenting with changing the budget decision structure in ways that appear to be redressing the traditional asymmetry of information in government and, in turn, giving elected decision makers in the executive and legislative branches the opportunity to more thoughtfully consider the public's demand for governmental goods and services. The final word is not yet in, nor is it likely to be in soon. By more strategically considering and testing the value of tenants of public choice theory and other fields to public budgeting, though, we are likely to move in the right direction.

NOTES

* The views expressed herein do not necessarily reflect the views of the U.S. General Accounting Office.

1. Weber may not have seen democratic administration as a long-lived viable alternative to bureaucratic administration (Ostrom, 1974), but a strong public choice advocate, Vincent Ostrom, interpreting the works of Hamilton, Madison and Tocqueville, and drawing on experiences of the French bureaucracy, did.

2. Wyckoff (1990, 35) clearly models the similarities and differences of budgetary maximizing bureaucrats and slack-maximizing bureaucrats. Under budget maximization, Wyckoff argues that the bureaucrat presents the legislature with an all-or-nothing budgetary choice, in effect an elastic demand curve. Therefore, to maximize the budget size, "a bureaucrat would have to reduce the price charged to the sponsor to the level of his costs, eliminating productive inefficiency." Consequently, no funds would be left for discretionary purposes, such as increasing staff size and increasing salaries, over what is needed to maintain productivity. If instead a bureaucrat is assumed to maximize slack, then more resources will be available to "purchase whatever non-productive expenditures the bureaucrat desires." Because of the mechanisms in place to control bureaucratic discretion, most agencies probably operate as part budget maximizing and part slack maximizing.

3. As Udehn (1996, 37) argues, sociologists variously "explain human action in terms of structural location and cultural belonging," not individual utility and self maximiza-

tion (also see discussions by Parsons (1967), Lazarsfeld et al. (1948), and Therborn (1991)).

4. Frey (1993, 97) argues: "In the future, the main emphasis should not lie on exporting economics but rather on *importing* aspects and insights from other broadly conceived social sciences. . . . What is needed, however, is an effort to overcome the model of 'homunculus economicus' who is at all times in full control of his or her emotions, who does not know any cognitive limitations, who is not embedded in a personal network, who is extrinsically motivated and whose preferences are not influenced by processes of discussion." And, he continues, "Experiments by psychologists . . . as well as by economists . . . have by now revealed overwhelming evidence that humans, as well as animals . . . do not act rationally in the sense of following the vonNeumann/Morgenstern axioms."

5. In this invited piece, Buchanan (1993, 68–69) argues that actions are determined not only by individual self-interests but by organizational structure as well. He argues that public choice theory has not paid enough attention to elementary principles of majority rule and majoritarian exploitation. He says, "The venality that is widely imputed to politicians and bureaucrats is surely misplaced in many instances, and concentration on overt misbehavior distracts attention from the organizational structure that allows the observed results to emerge . . . We do not necessarily get the results we see because politicians, like the rest of us, are sometimes motivated by self interest. We get the results we see because the incentive structures of politics insure the survival of those politicians who do depart from the public interest norms."

6. As suggested earlier, public choice advocates tend to explain government behavior from a very restricted disciplinary view. In addition to their general distaste for the social sciences generally (even though as Louis Weschler (1982, 294) argues, they are "a full member of social science"), Lovrich & Neiman (1984) argue that their research ignores much of the findings in urban affairs and narrowly addresses equity and the consequences of externalities.

7. Seidman, however, argues that many of the criticisms of orthodox views of public administration are very shallow. He continues to suggest that the principles of orthodoxy, in spite of their limitations, do represent guidelines that have been used to guide the policies of various presidential administrations, including the very important Reorganization Act of 1949.

8. In this chapter, especially on pages 283–286, Stevens looks at competing principal-agent models. Among the models discussed are iron triangles, agency dominance, and congressional dominance.

9. In this case, Stokey & Zeckhauser (1978, 281) point out that "Policy makers must in effect make interpersonal comparisons of welfare, and then make judgments about the way welfare should be distributed."

ACKNOWLEDGMENTS

I want to thank Rebecca Hendrick, Tom Lauth, the late Charles Spindler – one of my best friends whom I will remember for his friendship, sense of humor, and unparalleled professionalism – Irene Rubin and Fred Thompson for the insights they have given me on government accountability and budgetary reform. An anonymous reviewer and John Swain, as well as, of course, John Bartle who helped point out glaring errors and omissions made in earlier versions of this paper. I also cannot say enough for the enormous patience of my wife Debbie.

REFERENCES

Appleby, P. H. (1949). *Policy and Administration*. University AL: University of Alabama Press.

Arrow, K. J. (1951). *Social Choice and Individual Values*. New York: John Wiley and Sons.

Axelrod, D. (1995). *Budgeting for Modern Government* (2nd ed.). New York: St. Martin's Press, Inc.

Balbus, I. (1975). Politics as sport: An interpretation of the political ascendancy of the sports metaphor in America. In: L.N. Lindberg, R. Alford, C. Crouch & C. Offee (Eds), *Stress and Contradiction in Modern Capitalism* (pp. 321–336). Lexington MA: Lexington Books.

Behn, R. B. (1980). Leadership Tactics for Retrenchment. *Public Administration Review*, *40*, 604–608.

Bergson, A. (1938). A Reformulation of Certain Aspects of Welfare Economics. *Quarterly Journal of Economics*, *52*, 314–344.

Berman, L. (1979). *The Office of Management and Budget and the Presidency, 1921–1979*. Princeton NJ: Princeton University Press.

Borooah, V. K. (1996). Widening public choice. In: J. C. Pardo & F. Schneider (Eds), *Current Issues in Public Choice* (pp. 43–50). Cheltenham, U.K.: Edward Elgar.

Brady, G. L., & Tullock, G. (Eds) (1996). *Formal Contributions to the Theory of Public Choice: The Unpublished Works of Duncan Black*. Boston: Kluwer Academic Publishers.

Brenton, A., & Wintrobe, R. (1975). The Equilibrium Size of a Budget Maximizing Bureau: A Note on Niskanen's Theory of Bureaucracy. *Journal of Political Economy*, *83*, 195–207.

Buchanan, J. M. (1949). The Pure Theory of Government Finance: A Suggested Approach. *Journal of Political Economy*, *57*, 496–505.

Buchanan, J. M. (1966). An individualistic theory of political process. In: D. Easton (Ed.), *Varieties of Political Theory* (pp. 25–37). Englewood Cliffs NJ: Prentice Hall.

Buchanan, J. M. (1986). *Liberty, Market and State*. Brighton: Wheatsheaf Books.

Buchanan, J. M. (1988). The constitution of economic policy. In: J. D. Gwartney & R. E. Wagner (Eds), *Public Choice and Constitutional Economics* (p. 35). Greenwich CT: JAI Press, in Udehn.

Buchanan, J. M. (1989). *Explorations into Constitutional Economics*. College Station: Texas A&M University Press.

Buchanan, J. M. (1993). Public Choice After Socialism. *Public Choice*, *77*, 67–74.

Buchanan, J. M., & Tullock, G. (1965). *The Calculus of Consent: Logical Foundations of Constitutional Democracy*. Ann Arbor: University of Michigan Press.

Caiden, N. (1980). Budgeting in Poor Countries: Ten Common Assumptions Re-examined. *Public Administration Review*, *40*, 40–46.

Caiden, N., & Wildavsky, A. (1974). *Planning and Budgeting in Poor Countries*. New Brunswick NJ: Transaction.

Crozier, M. (1964). *The Bureaucratic Phenomenon*. Chicago: The University of Chicago Press.

Cyert, R. M., & March, J. G. (1963). *A Behavioral Theory of the Firm*. Englewood Cliffs NJ: Prentice-Hall.

Davis, O. A., Dempster, M. A. H., & Wildavsky, A. (1974). Toward a Predictive Theory of the Federal Budgetary Process. *The British Journal of Political Science*, *4*, 419–452.

De Borger, B., Kerstens, K., Moesen, W., & Vanneste, J. (1994). Explaining Differences in Productive Efficiency: An Application to Belgian Municipalities. *Public Choice, 80*, 339–358.

DeGregori, T. R. (1974). Caveat Emptor: A Critique of the Emerging Paradigm of Public Choice. *Administration and Society, 6*, 205–228.

Downs, A. (1957). *An Economic Theory of Democracy*. New York: Harper and Brothers.

Downs, A. (1967). *Inside Bureaucracy*. Boston: Little, Brown.

Duncombe, S., & Kinney, R. (1987). Agency Budget Success: How it is Defined by Budget Officials in Five Western States. *Public Budgeting and Finance, 7*, 24–37.

Dunn, W. N. (1994). *Public Policy Analysis: An Introduction* (2nd ed.). Englewood Cliffs NJ: Prentice-Hall, Inc.

Elkin, S. L. (1987). *City and Regime in the American Republic*. Chicago: University of Chicago Press.

Elster, J.(1986). The Market and the Forum: Three Varieties of Political Theory. *Economics and Philosophy, 1*, 231–265.

Fenno, R. F., Jr. (1966). *The Power of the Purse: Appropriations Politics in Congress*. Boston: Little, Brown and Company.

Forrester, J. P. (1993). The Rebudgeting Process of State Government: The Case of Missouri. *The American Review of Public Administration, 23*, 155–178.

Forrester, J. P., & Mullins, D. (1992). Rebudgeting: The Serial Nature of Municipal Budgetary Practices. *Public Administration Review, 52*, 467–473.

Forrester, J. P., & Adams, G. B. (1997). Budgetary Reform Through Organizational Learning: Toward an Organizational Theory of Budgeting. *Administration & Society, 28*, 466–488.

Frey, B. S. (1993). From Economic Imperialism to Social Science Inspiration. *Public Choice, 77*, 95–105.

Governmental Accounting Standards Board (1987). *Concepts Statement No. 1, Objectives of Financial Reporting*. Norwalk CT: GASB.

Governmental Accounting Standards Board (1994). *Concepts Statement No. 2, Service Efforts and Accomplishments*. Norwalk CT: GASB.

Golembiewski, R. T. (1977a). A Critique of 'Democratic Administration' and Its Supporting Ideation. *American Political Science Review, 71*, 1488–1507.

Golembiewski, R. T. (1977b). Observations on Doing Political Theory. *American Political Science Review, 71*, 1526–1531.

Goodsell, C. T. (1983). *The Case for Bureaucracy: A Public Administration Polemic*. Chatham NJ: Chatham House Publishers, Inc.

Gulick, L. H. (1937). Notes on the theory of organizations. In: L. Gulick & L. F. Urwick (Eds), *Papers on the Science of Administration* (pp. 1–45). New York: Institute of Public Administration, Columbia University.

Haveman, R. H. (1976). *The Economics of the Public Sector* (2nd ed.) (pp. 145–151). New York: John Wiley & Sons,

Hawkins, B. W. (1966). Public Opinion and Metropolitan Consolidation in Nashville. *Journal of Politics, 28*, 408–418.

Heap, S. H., Hollis, M., Lyons, B., Sugden, R., & Weale, A. (1992). *The Theory of Choice: A Critical Guide*. Oxford UK: Blackwell Publishers.

Heclo, H. (1977). *A Government of Strangers: Executive Politics in Washington*. Washington D.C.: The Brookings Institution.

Heclo, H. (1978). Issue network and the executive establishment. In: A. King (Ed.), *The New American Political System* (pp. 87–124). Washington D.C.: American Enterprise Institute.

Hyman, D. N. (1990). *Public Finance: A Contemporary Application of Theory to Policy*. Chicago: The Dryden Press.

Kaufman, H. (1976). *Are Government Organizations Immortal?* Washington D.C.: Brookings Institution.

Kiel, L. D., & Elliott, E. (1992). Budgets as Dynamic Systems: Change, Variation, Time, and Budgetary Heuristics. *Journal of Public Administration Research and Theory*, *2*, 139–156.

Lauth, T. P. (1988). Mid-Year Appropriations in Georgia: Allocating the 'Surplus'. *International Journal of Public Administration*, *11*, 531–550.

Lazarsfeld, P. F., Berelson, B., & Gaudet, H. (1948). *The People's Choice*. New York: Columbia University Press.

Levine, C. (1978). Organizational Decline and Cutback Management. *Public Administration Review*, *38*, 316–325.

Levine, C. H. (1979). More on Cutback Management: Hard Questions for Hard Times. *Public Administration Review*, *39*, 179–183.

Levine, C. H., Rubin, I. S., & Wolohojian, G. G. (1981). *The Politics of Retrenchment*. Beverly Hills CA: Sage.

Lindblom, C. (1959). The Science of Muddling Through. *Public Administration Review*, *19*, 79–88.

Logan, J. R., & Molotch, H. L. (1987). *Urban Fortunes: The Political Economy of Place*. Berkeley: University of California Press.

Long, N. E. (1949). Power and Administration. *Public Administration Review*, *9*, 257–264.

Lovrich, N. P., & Neiman, M. (1984). *Public Choice Theory in Public Administration: An Annotated Bibliography*. New York: Garland Publishing, Inc.

Marando, V. L., & Whitley, C. R. (1972). City-County Consolidation, An Overview of Voter Responses. *Urban Affairs Quarterly*, *8*, 181–203.

March, J. G. (1978). Bounded Rationality, Ambiguity, and the Engineering of Choice. *The Bell Journal of Economics*, *9*, 587–608.

March, J. G., & Olsen, J. P. (1976). *Ambiguity and Choice in Organizations*. Bergen: Universitets Forlaget.

McCaffery, J. (1988). *Budgetmaster* (2nd ed.).

McKenna, C. K. (1980). *Quantitative Methods for Public Decision Making*. New York: McGraw-Hill Book Company.

Meltsner, A. J. (1971). *The Politics of City Revenue*. Berkeley, CA: University of California Press.

Meyers, R. T. (1994). *Strategic Budgeting*. Ann Arbor MI: The University of Michigan Press.

Migué, J-L., & Bélanger, G. (1974). *Toward a General Theory of Managerial Discretion*. Public Choice 17, 27–43.

Mikesell, J. L. (1991). *Fiscal Administration: Analysis and Applications for the Public Sector*. Pacific Grove CA: Brooks/Cole Publishing Company.

Mintzberg, H. (1979). *The Structuring of Organizations*. Englewood Cliffs NJ: Prentice-Hall, Inc.

Molotch, H. (1976). The City as a Growth Machine: Toward a Political Economy of Place. *American Journal of Sociology*, *82*, 309–332.

Mueller, D. C. (1989). *Public Choice II: A Revised Edition of Public Choice*. Cambridge: Cambridge University Press.

National Performance Review (September, 1993). Mission driven, results-oriented budgeting. Office of the Vice President. Washington D.C. http://www.npr.gov/library/nprrpt/annrpt/sysrpt93/mission.htm

Niskanen, W. A. (1971). *Bureaucracy and Representative Government*. Chicago: Aldine-Atherton.

Niskanen, W. A. (1974). Comment. *Public Choice*, *17*, 43–44.

Niskanen, W. A. (1975). Bureaucrats and Politicians. *Journal of Law and Economics*, *18*, 617–643.

Olson, M. (1965). *The Logic of Collective Action: Public Goods and the Theory of Groups*. Cambridge MA: Harvard University Press.

The Oregon Option. (1994, July 25). www.npr.gov/library/fedstat/2642.html

Osborne, D., & Gaebler, T. (1992). *Reinventing Government: How the Entrepreneurial Spirit is Transforming the Public Sector*. Reading Mass.: Addison-Wesley Pub. Co.

Ostrom, E. (1971). Institutional Arrangements and the Measurement of Policy Consequences: Applications to Evaluating Police Performance. *Urban Affairs Quarterly*, *6*, 447–475.

Ostrom, V. (1974). *The Intellectual Crisis in American Public Administration* (revised ed.). University AL: The University of Alabama Press.

Ostrom, V. (1977). Some Problems in Doing Political Theory: A Response to Golembiewski's 'Critique.' *American Political Science Review*, *71*, 1508–1525.

Ostrom, V., & E. Ostrom. (1971). Public Choice: A Different Approach to the Study of Public Administration. *Public Administration Review*, *31*, 203–216.

Pareto, V. (1927). *Manuel D'economie Politique* (2nd ed.). Paris: M. Giard.

Parsons, T. (1967). *Sociological Theory and Modern Society*. New York: The Free Press of Glencoe.

Pearson, N. M. (1980). The budget bureau: From routine business to general staff. In: A. Schick (Ed.), *Perspectives on Budgeting* (pp. 138–143). Washington, DC: American Society for Public Administration.

Redford, E. S. (1969). *Democracy in the Administrative State*. New York: Oxford University Press.

Reinvention's Next Steps: Governing in a Balanced Budget World. Background Papers Supporting a Speech by Vice President Al Gore. (1996, March 4). Accessed on the World Wide Web at www.npr.gov/library/papers/bkgrd/balbud.html

Rowley, C. K., Schneider, F., & Tollison, R. D. (1993). The Next Twenty-five Years of Public Choice. *Public Choice*, *77*, 1–7.

Rubin, I. S. (1985). *Shrinking the Federal Government: The Effect of Cutbacks on Five Federal Agencies*. New York: Longman.

Rubin, I. S., & Rubin, H. J. (1987). Economic Development Incentives: The Poor (Cities) Pay More. *Urban Affairs Quarterly*, *23*, 37–62.

Rubin, I. S. (1990). *The Politics of Public Budgeting: Getting and Spending, Borrowing and Balancing*. Chatham NJ: Chatham House Publishers, Inc.

Rubin, I. S. (1993). Who Invented Budgeting in the United States? *Public Administration Review*, *53*, 438–454.

Ruggiero, J., Duncombe, W., & Miner, J. (1995). On the Measurement and Causes of Technical Inefficiency in Local Public Services: With an Application to Public Education. *Journal of Public Administration Research and Theory*, *5*, 403–428.

Samuelson, P. A. (1947). *Foundations of Economic Analysis*. Cambridge: Harvard University Press.

Schick, A. (1966). The Road to PPB: The Stages of Budget Reform. *Public Administration Review*, *26*, 243–258.

Schick, A. (1988). An inquiry into the possibility of a budget theory. In: I. S. Rubin (Ed.), *New Directions in Budget Theory* (pp. 56–69). Washington DC: The Urban Institute.

Schier, S. E. (1992). *A Decade of Deficits: Congressional Thought and Fiscal Action*. Albany: State University of New York Press.

Schultze, W. A. (1985). *Urban Politics: A Political Economy Approach*. Englewood Cliffs NJ: Prentice Hall.

Simon, H. (1945). *Administrative Behavior*. New York: Macmillan Publishing Co.

Simon, H. (1976). *Administrative Behavior* (3rd ed.). New York: Free Press.

Seidman, H. (1980). *Politics, Position & Power: The Dynamics of Federal Organization* (3rd ed.). New York: Oxford University Press.

Spindler, C. J., & Forrester, J. P. (1993). Economic Development Policy: Explaining Policy Preferences Among Competing Models. *Urban Affairs Quarterly*, *29*, 28–53.

Stevens, J. B. (1993). *The Economics of Collective Choice*. Boulder: Westview Press.

Stillman, R. J. (1976). Professor Ostrom's New Paradigm for American Public Administration – Adequate or Antique? *Midwest Review of Public Administration*, *10*, 179–192.

Stillman, R. J. (1978). Controversy in Public Administration – A Reply to Professor Ostrom. *Midwest Review of Public Administration*, *12*, 41–44.

Stokey, E., & Zeckhauser, R. (1978). *A Primer for Policy Analysis*. New York: W.W. Norton & Company.

Stone, C. N. (1980). Systemic Power in Community Decision Making: A Restatement of Stratification Theory. *American Political Science Review*, *74*, 978–990.

Stone, C. N. (1989). *Regime Politics: Governing Atlanta, 1946–1988*. Lawrence: University Press of Kansas.

Stone, C. N., & Sanders, H. T. (Eds) (1987). *The Politics of Urban Development*. Lawrence: University Press of Kansas.

Stone, C. N., Whelan, R. K., & Murin, W. J. (1986). *Urban Policy and Politics in a Bureaucratic Age* (2nd ed.). Englewood Cliffs, NJ: Prentice-Hall.

Straussman, J. D. (1988). Rights-based budgeting. In: I. S. Rubin (Ed.), *New Directions in Budget Theory* (pp. 100–124). Albany NY: SUNY Press.

Straussman, J. D., & Thurmaier, K. (1989). Budgeting Rights: The Case of Jail Litigation. *Public Budgeting and Finance*, *9*, 30–42.

Therborn, G. (1991). Cultural Belonging, Structural Location and Human Action: Explanation in Sociology and in Social Science. *Acta Sociologica*, *34*, 177–191.

Tullock, G. (1967). *Toward a Mathematics of Politics*. Ann Arbor: University of Michigan Press.

Tullock, G. (1970). *Private Wants, Public Means: An Economic Analysis of the Desirable Scope of Government*. New York: Basic Books, Inc.

Tullock, G. (1992). *Economic Hierarchies, Organization and the Structure of Production*. Boston: Kluwer Academic Publishers.

Tullock, G., & Wagner, R. E. (Eds) (1978). *Policy Analysis and Deductive Reasoning*. Lexington: D.C. Heath and Company.

Udehn, L. (1996). *The Limits of Public Choice: A Sociological Critique of the Economic Theory of Politics*. London: Routledge.

van Winden, F. A. M. (1988). The economic theory of political decision-Making: A survey and perspective. In: VanDenBroeck, J. (Ed.), *Public Choice* (pp. 9–42). The Netherlands: Kluwer Academic Publishers.

Waldo, D. (1948). *The Administrative State*. New York: The Ronald Press Co.

Waldo, D. (1952). Development of a Theory of Democratic Administration. *American Political Science Review*, *46*, 81–103.

Weschler, L. F. (1982). Public Choice: Methodological Individualism in Politics. *Public Administration Review*, *42*, 288–294.

Wicksell, K. (1896). A New Principle of Just Taxation. Finanztheoretische Untersuchungen. Jena, reprinted in R. A. Musgrave & A. T. Peacock (Eds) (1967). *Classics in the Theory of Public Finance* (pp. 72–118). New York: St. Martin's Press.

Wildavsky, A. (1988). *The New Politics of the Budgetary Process*. Glenview IL: Scott, Foresman.

Wyckoff, P. G. (1990). The Simple Analytics of Slack-Maximizing Bureaucracy. *Public Choice*, *67*, 35–47.

7. THE TRUTH IS OUT THERE: IS POSTMODERN BUDGETING THE REAL DEAL?

Janet Foley Orosz

INTRODUCTION

My work on this chapter evolved into the pursuit of what White (1999, 178) urges is a responsibility of those of us who are engaged in the academic study of public administration. That is, we should pursue self-reflection on the local narratives with which we define our academic communities and especially continue conversations about "our historical understanding of the field, the cultural norms to which we adhere, the languages we speak, and our willingness to talk to each other." I hope to enter developments in public administration theory into the local "conversational community" (White, 1999, 177) of public budgeting.

When I was asked to write this chapter on alternative approaches to budgeting, I planned to concentrate on the potential for interpretive or constructivist views of budget processes, a general grouping of approaches that are alike in their relativistic ontologic position (Guba & Lincoln, 1994, 109). I also wanted to include comments on the value of developing a critical theory of budgeting. Here again, I mean to use the classification "critical budget theory" as an umbrella term. Put simply by Guba and Lincoln (1994, 109) critical theory is "a blanket term . . . denoting a set of several alternative paradigms . . . [which] may be divided into three substrands: post-structuralism, postmodernism, and a blending of the two. Whatever their differences, the common breakaway

Evolving Theories of Public Budgeting, Volume 6, pages 125–155.

ISBN: 0-7623-0790-0

assumption of all of these variants is that of the value-determined nature of inquiry – an epistemological difference."

In the years(!) that passed while I worked on other projects, public administration theorists, reflecting on the status of knowledge in public administration, began to write about the potential for postmodernism and post-structuralism to inform examinations of the public administration field. Specifically, some public administration theorists embrace the postmodern era – one in which the metanarrative and implications of technical rationality as a framework (the modern era) are dissected, the constructions that underlie them revealed, and their implications identified (Adams, 1992; Farmer, 1995, 1996; Fox & Miller, 1995, 1997; White & Adams, 1994).[1] This turn presented me with a new opportunity to use this chapter to examine the construction of language and narratives of public budgeting, what is sometimes termed a poststructural analysis.[2] My effort follows in the largely unnoticed footsteps of Gerald Miller. His (1991) book *Government Financial Management Theory* is rarely cited but is path-breaking in its focus on the constructions of public financial management and budgeting.

In this chapter, I review some of the existing narratives of public budgeting, in theory and in professional practice. I write from the standpoint of a public administration theorist and a teacher/ practitioner of public budgeting, rather than as a budgeting theorist who has been an integral participant in the construction of academic public budget theory. This chapter expands consideration of public budgeting to include interpretive and critical, and some postmodern narratives in addition to the dominant dialogues in both the practice and academic study of public budgeting. Since these approaches have not been the norm in public budgeting and finance, this chapter has a limited number of examples of published work from alternative budget approaches. My coverage of budget narrative history serves to direct attention to the norms and local narratives that influence the development of – or inattention to – alternative forms of knowledge about public budgeting.[3]

In his book *Reflections on the Growth of Knowledge in Public Administration*, Jay D. White (1999, 159) observed that "Specialization of knowledge has resulted in the creation of many local narratives or language games, which prevents meaningful discourse across them and sometimes even within them." White goes on to say that "Researchers from different perspectives run the risk of talking past one another because of their local narrative or language game, thus failing to achieve a domain of practical discourse. A warfare of sorts can erupt when researchers from one perspective disregard out of hand the stories told by researchers from other perspectives."

This volume on budgeting and financial management in the *Research in Public Administration* series, and its component chapters, is itself an example of bridging local narratives and continuing dialogues across local narratives within budgeting and public administration. At the Association for Public Budgeting and Financial Management in October 1999 in Washington, D.C., three of the authors of chapters within this book were on a panel. It was obvious to me, and probably to all in the room, that a clearly different narrative underlies my work on this chapter. Frankly, there wasn't much exchange across the panelists about their work (different languages). To the credit of the panel's discussant, the submitted works were treated as contributions based on different assumptions; economic modeling, "barefoot empiricism" and interpretive hermeneutics. The discussant suggested budget theorists might take a pragmatic perspective in selecting between approaches for a given research topic.[4] Including a chapter on alternative budget theory in this book is a signal that meaningful conversation is possible between theorists operating from different narratives, with the purpose of advancing our understanding of public budgeting processes and theories.

AN APPLIED EXAMPLE OF INTERPRETIVE AND CRITICAL APPROACHES TO BUDGETING

I will eventually proceed with more theoretical and research-related accounts of the differences between interpretive, critical, and some postmodern approaches to public budgeting and finance. Particularly, I will address how these approaches relate to existing and multiple structural assumptions in public budgeting currently held by practitioners and academics. Before getting into this, the following case example from my own practitioner experience may help to show the differences between budgeting from technical-rational, interpretive, and critical approaches.

Budget cuts and central support charges in the Ohio Department of Natural Resources

The early 1980s produced hard times for Ohio's state government finances. State budgets absorbed the impact of federal cuts from the mini-recession, and income tax collections did not support the spiraling costs of medicaid and welfare benefits. During this time, some Ohio state agencies absorbed across-the-board and some selective cuts so that the state could meet its legal obligations in the midst of the biennial budget. When the next biennial budget was prepared, officials and analysts in the Ohio Office of Budget and

Management (OBM) were determined to have a more rational approach to decreasing agency reliance on general revenue funds, and OBM performance incentives for analysts reflected this objective. Budget development and adoption processes in 1982–1983 coincided with the change in Ohio's Governor and Governor's political party. The new Governor wanted general revenue funds freed up for his own new initiatives. This case of the Ohio Department of Natural Resources' (ODNR) funding for central support offices (purchasing, personnel, director's office, public information, real estate services, engineering, business and finance functions) exemplifies the shifting of support away from general revenue funds to other funding sources, especially in the case of agency divisions that had stable revenue sources from fees and licenses.

Specifically, OBM proposed the establishment of a chargeback system that required divisions of Wildlife and Watercraft (as well as other non-general revenue-funded Divisions) to pay for the central support services based on cost-accounting principles. The Department already utilized standardized federal cost accounting principles to determine an overhead charge for federal grants, and this principle was to be transferred to other non-general revenue fund accounts that received central support services. The projected general revenue funds to be freed-up for other uses was $7.0 million (Griffith & Associates, 1984), and this amount was removed from the Department's general revenue fund allocation in the fiscal year 1984 executive budget and replaced with additional appropriation authority in the Department's non-general revenue fund accounts.

From the beginning, the central support chargeback system, as it came to be called, was resisted by ODNR officials of both the outgoing and incoming administrations. OBM analysts and supervisors insisted, however, that the system be implemented and general revenue support was removed from the budget. In this approach, "proxies" for services of each office were established (for example, number of invoices processed for business and finance, or number of personnel descriptions written by employee services). The Division of Wildlife, a heavy user of central support services, was assigned the largest share of costs to be recovered, $2.7 million. Wildlife advocacy groups saw this as a direct reduction in program and operating funds generated from the fees and licenses legally required to be used for specified purposes by the Wildlife Division. To these groups, chargebacks for central support meant more of their money was diverted to departmental administration. Further, if the chargeback system was to be accurate, then the system was too complex to be explained. On the other hand, simplicity in chargeback mechanisms resulted in unreasonable-looking costs to process invoices or vouchers. ODNR's Assistant Director and Budget Officer developed a dog-and-pony show that was delivered to wildlife constituent groups and sportswriters who covered Division of Wildlife activities in the hopes of explaining the system and shifting "blame" to the Office of Budget and Management. The chargeback system, although implemented and adjusted, was never accepted by the Division of Wildlife and other Division proponents in the legislature. A member of the ODNR administrative team making budget and policy decisions for the new gubernatorial administration later commented:

> . . . first of all, I don't think we understood the consequences of it probably enough. And second, I think, we didn't think it serious because we thought once the budget gets in the legislature, this nonsense will cease. And it didn't. I think more than anything else, that was my greatest frustration at the Department. I spent more time trying to figure out ways to implement and go forward with the chargeback system than I did with policy issues . . . I bet in some weeks 50% of the time was devoted, directly or indirectly, with trying to deal with that [chargeback system] (Napier, 1989).

Although the chargeback system continued through two biennial budgets, it remained an issue at budget time. In a subsequent budget bill, the system was abolished and general revenue funds were allocated to pay back the Wildlife Fund, with interest.

From the traditional accounting/budget metaphor of technical rationality held by the Office of Budget and Management analysts in the above example, a move to a central support chargeback system in which non-general-revenue-funded divisions paid their way made perfect sense. However, in the socially-constructed world of agency management, "chargeback" took on a life of its own, requiring inordinate amounts of time on the part of managers on all sides of the proposal. The accounting/technical rationality metaphor of the budget office could not be forced onto the situation, although it did frame the initial dialogue in the Office of Budget and Management and ODNR officials. The metaphor of paying a fair share was shifted by ODNR stakeholder groups to a metaphor defined by the misuse of Ohio sportsmen's cash, cash that was legally restricted to supporting hunting and fishing purposes. Although related budget changes could possibly be identified through analysis of memos and official and unofficial budget documents, this story could only partially be understood through this type of technical analysis. The full social construction of the central support chargeback system could only be known through discussions with multiple participants – an interpretive approach to understanding budgeting.

The case could also be considered from a critical perspective by asking these questions: What were the interpretations that led to linking the chargeback of central support costs (and how they were defined) to the biennial budget? What was the power situation that produced this chargeback system and the opposition toward it? What was attention diverted from when the language of budget issues focused on central support chargebacks and cost accounting mechanisms? Who was privileged and who was excluded from the policy formation? What outcomes would have resulted from an alternative model? These types of questions form the basis for a critical approach to budgeting.

UNDERSTANDING ALTERNATIVE APPROACHES TO KNOWLEDGE IN PUBLIC ADMINISTRATION AND USING ALTERNATIVE APPROACHES TO KNOWLEDGE TO EXPAND BUDGET THEORY

Having taken a look in this small way at the value of including interpretive and critical approaches to public budgeting as making a contribution to understanding the real life experiences of those involved with public budgeting, I am advocating the need for activity in these alternative approaches in the study (theory) of public budgeting. While most budget research is conducted using the analytic approaches prevalent in modernist social sciences, this analytic system can be viewed as a construction in itself, both a lived system constructed by social structures based in technical rationality, and a construction of the research community (Fox & Miller, 1997; Adams, 1992; White & Adams, 1995; White, 1999). In White's words (1999, 167):

> As a profession, public administration has its own norms and values, that, along with beliefs, guide professional actions. The same can be said for the academic study of public administration. Academicians have their own beliefs about what should be studied in public administration and how it should be studied. Academicians also have shared beliefs, norms, and values that guide the practice of their research. Thus the professional practice and academic study of public administration are structural phenomena in society. These structures are not, however, universal or invariant. Actually they are local and also multiple.

Traditionally, while some researchers involved in public budgeting may focus on the people making policy and budget choices in a social world, their aim in research is on discovering what is happening "out there." For example, in examining the politics of the budget decision process, researchers and theorists focused on identifying and reporting on a verifiable truth rather than on the social and organizational impacts of budget decisions and processes (Wildavsky, 1992; Rubin, 1993; Thurmaier & Gosling, 1997). Farmer (1996, 131) elaborates on this modernist approach:

> ... modernism has been optimistic about the power of human reason. It accepts the non-relativistic view that observations can be made – and conclusions can be developed – independent of the lens of the observer. The reflexive or perspectival view, discussed by Derrida and before him by Nietzshe, Heidegger, and Wittgenstein, takes us to the post-modern world. This is a world recognizing that there are no certainties, no context-free truths; there is not even the certainty represented by this claim.

Also, in the modernist approach, there is a limited critique of the larger, deeper structured context that produced these experiences. When applied to the realm of public budgeting and financial management, postmodern theorists would urge reflective consideration of the conditions that produced acceptance and adoption (formally or informally) of the approaches to budgeting that are taught and practiced today. Farmer (1995, 1996) labeled this the reflexive language paradigm," [one that] "recognizes the value of identifying how the object (or fact) seen is socially constructed, in part at least, by the theoretical lens or perspective utilized."

Interpretive approaches, critical theory and postmodern theory are linked through their reflective and philosophical bases (Farmer, 1995, 1996, 1997; Fox & Miller, 1995; Jun & Rivera, 1997; McSwite, 1997; Marshall & Choudhury, 1997; White, 1999; Woller, 1997; Woller & Patterson, 1997).[5] Postmodern theory seemingly has its own language,[6] emphasizing deconstruction and reflexivity. To explain these terms and their relation to postmodern approaches, I rely on David Farmer's (1996, 129) description from his work that extends postmodernism to public administration theory:

> Deconstruction is indeed the act of close reading and of breaking through the defenses of a text (where text includes not only the documents but also lives and actions), showing that privileged terms depend on other even more significant excluded terms; it is the act of X-raying a picture (situation) to reveal another picture (situation).

Farmer (1996, 131) continues with an explanation of the postmodern term "reflexivity":

> ... knowledge [about public administration] is always relative to a perspective. This reflexive or perspectivist epistemological position, embraced by postmodernism, amounts to holding that we are trapped in a cobweb of perspectives, and that our understanding of the real (i.e. truths) is limited to the perspective of a particular theoretical frame – a frame shaped by the observer's context of knowledge, prejudices, assumptions, ideas, intuitions, and desires. This does not entail, however, that each perspective is equally good – or even satisfactory.

This is the same claim that Gerald Miller, working from an interpretive and critical framework, addressed in his book *Government Financial Management Theory* (1991). The examination of budgeting and financial management produced by Gerald Miller stands as an early exemplar of a structural analysis done from a postmodern framework. In this work, Miller (as did Forrester and

Adams in 1997) suggests that the socially-constructed world of public budgeting can be addressed from an organizational theory perspective rather than or in addition to an economic or technical one.[7] The Ohio Department of Natural Resources central support case described above allowed us, at some minimal level, to practice reflexivity and deconstruction of a specific situation, to see that there are alternative interpretations of budget policy circumstances.

I now turn to accounts of current changes in public administration theory away from the rational choice models and integrate more detailed accounts of related efforts in public budgeting.

NARRATIVES OF PUBLIC ADMINISTRATION: PUBLIC BUDGETING AND FINANCIAL MANAGEMENT PARALLELS

Timney (1995), in what she termed a "synchronism" argument, observed that while public budgeting and public administration were at one time aligned in their direction and emphasis, public administration theory is moving (although not uniformly) into the postmodern era beyond the framework of technical rationality, while public budgeting and finance remain primarily in the economic/rational or political frameworks. Just as momentum re-builds for public administration theorists to examine the underlying structures of the field,[8] periodic appraisals of developments in budget theory have also occurred, however, with less consideration of the underlying structures of budgeting narratives and more with the intent of evaluating the status and future direction of budget theory. Timney, as well as Miller (1991), and Farmer (1996), suggests that public budget theory would likewise benefit with inclusion of these alternatives as legitimate approaches to knowledge in the realm of public budgeting. This expansion, in public administration and budgeting, is not without its influential critics.[9]

Existing Narratives

According to public administration theorists White and Adams (1995, 4) and White (1999, ir) there are at least six local narratives of public administration. As Timney indicated in her syncronism argument, these narratives also play out in the history and practice of public budgeting, but are in many cases not inclusionary or sequential. These narratives of public administration are listed in Table 1 to which has been added parallel developments in the subfield of public budgeting and financial management.

Other budgeting/public administration theorists have offered their informed assessments of the past, present, and future states of public budgeting theory in the United States. These assessments include historical approaches that consider public budgeting trends and reforms, their usefulness, the explanatory and predictive validity of budget theories, their fit with practice (Rubin, 1990), reliance on technical rationality (Miller, 1991; Timney, 1995; Schick, 1966) and relationship to democratic theory and politics (Key, 1940; Rubin, 1998, 1994; Timney, 1995; Wildavsky, 1992; Alexander, 1999). I utilized these and other accounts as I considered how to place various budget initiatives within the corresponding narratives of public administration.

According to White (1999) and White and Adams (1995), a *first* narrative of public administration revolves around the Constitutional separation of powers and its relationship to bureaucratic responsibilities and structures. In the sub-field of public budgeting, procedures establishing legislative responsibilities in budget development, the executive budget process and budget administration (including structures of accountability and structures creating constraints on executive spending) each rest in the narrative of the constitutional separation of powers.

A *second* narrative, the dichotomy of politics and administration (Wilson, 1887), produced the image of the politically neutral and technically competent public administrator, and was accompanied by the *third* narrative, the scientific study and practice of public administration (Gulick & Urwick, 1937). Timney (1995) and Rubin (1990), among others, link public budgeting development in this period with classical public administration's emphasis on efficiency and value-free expert systems, with rationality and a minimum of political involvement on the part of the budget practitioner.

When practitioners and theorists alike refuted the dichotomy between politics and administration, the astute budget practitioner was urged (and budget theorists urged by practitioners) to understand that budget-making is political and that effectiveness in budget and finance involves understanding and operating within rather than outside that political environment. Still, efforts were made to develop systems that infused technical-rational information to political decision makers (Pyhrr, 1973). Wildavsky (1964, 1992) on the other hand, argued that budgeting was essentially about politics, and that national political stability was achieved through incrementalism.[10] Both types of observations were based in descriptive examinations of gaps between the theory and practice of budgeting (Rubin, 1990).

The *fourth* narrative of public administration identified by White and Adams centers on the relationship of theory and practice – whether and how theory informs practice and whether and how practice informs theory. Extending this

Table 1. Public Administration and Budgeting Narratives.

	Public Adminsistration[1] Narratives	Budget/Financial Management Parallel Narratives
Narrative 1	Constitutionalism	Accountability an auditing Budget systems Budget processes – legistrative/ executive balance
Narrative 2	Dichotomy of politics and administration	Budgeteers and finacial managers as implementors vs. Budget development as policymaking
Narrative 3	Scientific study and practice of public administration	Application of universal analytic techniques Generalization as goal Efficiency as applied goal
Narrative 4	Theory informing practice and practice informing theory	Politics/administration dichotomy as false Incrementalism Technical budget reforms
Narrative 5	The new public administration: Equity-focused; Citizen participation Rejection of positivism	Budget as a policy tool Citizen input in local budget processes Interest groups/stakeholders as participants
Narrative 6	Feminist theory Gender as a lens to view public administration history	None identified
Narrative 7	The new public management The business of government	Analysts as policy makers (valuing rational analysis) Efficiency as standard for service continuation Citizen as consumer
Emergent narrative	Postmodern, Alternative structures, Multiple and local narratives	Gerald Miller (1991) Government Financial Management Theory

[1] Modified from J. D. White & G. B. Adams (1995). Making Sense with Diversity: The Context of Research, Theory and Knowledge Development in Public Administration. In: *Research in Public Administration*. Newbury Park, CA: Sage. Related material also appears in J. D. White (1999). *Taking Language Seriously: The Narrative Foundations of Public Administration*. Washington D.C.: Georgetown Press.

narrative into budgeting, the theory/practice gap experienced by budget practitioners in the implementation of budget reforms led to adjustments in budget theory (Lauth, 1978; Schick, 1978; Wildavsky, 1964, 1986, 1992).[11] Many of these insights were based upon case studies and field work. These reforms shifted to program-related objectives contained in budgets (program budgets), eliminated the assumption of continuing and base line funding levels (zero-based budget reforms), or established expenditure ceilings through departmental allocation targets or "envelopes." Real time budgeting reforms were suggested as a budget form that might offer more discretion to program administrators in the field (Rubin, 1991). However, each of these reforms remain based in the model of technical rationality and exists largely within the constraints of existing budgeting systems (Adams, 1992).

The *fifth* narrative identified by White and Adams was at its height around the time of a 1968 conference of public administration scholars held at Syracuse University's Minnowbrook Conference Center. This narrative emerged from Minnowbrook with the label "the new public administration." The fifth narrative contains an emphasis on equity, citizen participation and rejection of positivism. In the "new PA," public administration theorists urged public administrators to embrace values beyond economic efficiency and to advocate for clientele (Marini, 1971).[12] Some theorists in this new tradition took a critical look at the implications of administrative actions on clients, and on the role of the administrator vis-a-vis clients (Harmon, 1981, 1995; Hummel, 1977).

For the most part, executive budgeting, an act which was far removed from clients, did not reflect the client-centered, equity-based orientation of the new public administration. Rather, the heart and soul of budgeting and financial management remained economic efficiency, accounting, and technical skills in budgeting[13] (Ammons, 1991; Axelrod, 1995; Mikesell, 1995). With varying degrees of authenticity, citizen participation was included in some budget procedures, especially at the local level. Values were not addressed as a legitimate part of technical budget construction – measures of efficiency prevailed with budget makers and implementors.

Moving out of this fifth narrative, in the 80s and 90s some public administration theorists "attacked the bases of bureaucracy, hierarchy and positivism" and considered its impact on an increasingly diverse workforce, as well as renewed concerns about connections and relationships with citizens (Timney, 1995, 46–47).[14] Solutions advocated included changes in public organization structures to flexible, responsive, non-hierarchical organizations (Harmon, 1981), government that is responsive to and part of the needs of a diverse citizenry, (King & Stivers, 1998) and an awareness of those who are ordinarily excluded from acts and accounts of public administration (Stivers, 1993). White and Adams describe the *sixth* narrative as feminist theory, the use of gender as a lense through which to view public administration history. This narrative appears to be virtually non-existent in the subfield of public budgeting and finance.

An evolving research agenda advises that state budget analysts in executive agencies are expanding their influence into agency policy-making through budget-related decisions (Gosling, 1987; Thurmaier, 1995; Thurmaier & Gosling, 1997; Willoughby, 1993) and describes budget analysts participating in the politics of budgeting through contact with the elected executive (albeit at the analyst level this might mean offering "analytic information" for consideration by political officials).[15] This shift accompanies what White and Adams (1995) term the "new public management," or *seventh* narrative of public administration (Lynn, 1996). Each of these narratives, with the exception of the sixth narrative of gender, remains based upon the technical-rational "tacit grand narrative of public administration" (Adams, 1992).

Developing Alternative Approaches to Knowledge in Public Administration and Public Budgeting

In this section, I build on the descriptions of alternative forms of knowledge that I presented earlier in the chapter. Starting in the late 1980s and gaining force through the 1990s, some public administration theorists again began to call for inclusion of interpretive, critical, and postmodern approaches to the study of public administration in addition to public administration's emphasis on technical rationality as an underlying framework for public administration (Adams, 1992; Denhardt, 1984; Fischer, 1981; Harmon, 1981; Orosz, 1997, 1998; White, 1986b, 1999; White & Adams, 1994; Zanetti, 1997, 1998). I labeled this the "emerging paradigm" in Table 1.

Although postmodern and critical accounts of budgeting are limited, there are examples of postmodern approaches in public policy that exemplify how public budgeting theory might look if approached from a postmodern stance (Chock, 1995; Colebatch, 1995; Linder, 1995; Pal, 1995; Yanow, 1996; Fischer & Forester, 1993; Kelly & Maynard-Moody, 1993). For example, Helen Liggett (1997) brought a postmodern perspective to the administration of Community Development Block Grants (CDBG).[16] Liggett's approach is syncopatic with the framework contained within Fox and Miller's (1995) *Postmodern Public Administration.* According to Fox and Miller (1997), postmodern public administration roots its analysis of public events and policy in four stages of consideration (representational, skeptical, radical absence, and hyper-reality). In Liggett's case example, CDBG allocation rules produced a radical absence of consideration of the issues underlying poverty. She stated, "What was definitely not required of policy analysts and planners involved in administering CDBG monies in community development agencies was an analysis of the causes of slum and blight and a plan for eliminating them . . .

Rather, the policy focused on qualification as a slum or blighted area" (Liggett, p. 20, emphasis added). Liggett demonstrated how politics and values relating to technical rationality entered the policy analysis process, and identified how "technical rationality" obscured structural and political issues that underlie the technical questions.

Approved Scripts of Academic Public Budgeting and Financial Management

Even though the inclusion of new narratives is occurring slowly in public administration and public policy analysis, public budgeting and accounting systems still remain more firmly embedded in the technical-rational accountability framework. There are several likely reasons for the slow development of interpretive and critical approaches to public budgeting and financial management. Some reasons are practical in origin. At the applied level, the budget cycle drives the work lives of many administrators in federal, state, and local governments. Annual appropriations processes shape program schedules, spending limits, and types of spending. Vacations and workloads are scheduled around budget cycle requirements. Students of introductory public administration might be asked to define the term "budget cycle," as though it is a real thing and not an artifact of the technical and legal structures created over the years for purposes of government accountability and control. Applied budgeting as a sub-field of public administration is lodged in a relatively easy reliance on the numbers that are involved in budgeting and the entrenched and demanding schedule of annual budgeting – the budget cycle.

Resource allocation decisions, limits, and possibilities, both overt and hidden, lie within the structure of the annual/biennial budget and related accounting systems. Accounting requirements drive the structure of expenditure information for public programs (Mikesell, 1995; Miller, 1991). Accounting and economics and their numbers-related approaches also remain dominant in the applied field of public budgeting because of reinforcing legal structures of society reflected in requirements for balanced budgets at state and local levels, reports and other data requirements of financial institutions necessary to pursue a favorable bond rating, and financial year-end reporting necessary for reasons of public accountability. As noted by budget reformers, while accounting structures, and the related focus on the numbers provide politicians and citizens closer control over agency activities, they do not necessarily monitor or enhance effectiveness in achieving agency policy goals, or in the parlance of economics, the efficient allocation of resources.

Finally, this technical-rational narrative is reinforced and self-replicating through budget education and on-the-job training. Cope (1989) acknowledges that there is an implicit, if not formally recognized, theory of executive budgeting that is taught in MPA schools, and that is used in governmental budget offices.[17] Much of the budget and financial management course work offered in schools of public policy and management embraces the narrative of scientific study and analytic practice of public administration.[18] Farmer (1996, 132) described conditions for the study of public administration and what counts as knowledge in public administration:

> Foucault's account of knowledge emphasizes normalization and power. Normalization implies the development and inculcation of whatever knowledge is required in order to produce the kind of person and the kind of behavior that is wanted . . . Foucault holds that what counts as truth, as real, is a reflection of the power situation in society. Similarly, do not public administration programs tend to count as real knowledge that knowledge which will assist graduates to fit into the work force?

Although not using labels associated with postmodern theory, Gerald Miller (1991, 9) took a deconstructionist/critical look at whose interests in academe are furthered by prevailing theories in public financial management, "including the topics chosen, the methods used, the explanations given, and the solutions to problems found." He suggests the prevailing theories are just one way of looking at financial management in the public sector and that there is room and value to alternative approaches. Miller's argument parallels that of White (1992) and White and Adams (1995) in public administration in the 1990s, as well as Forrester and Adams's (1997) thoughts that are specific to the organizational implications of budgeting. All argue for the acceptance of alternative approaches to knowledge.[19] Fox and Miller (1997, 86–87) offer this commentary on reasons for the trickling of support shown for changes in public administration curricula and the implied acceptance within the field.

> This necessity [of context specific, ideographic knowledge] seems not to have been emphasized in the public administration curriculum, perhaps because it stands in direct contradiction to a social science whose project is to make generalizable statements that are true across situations, that is, across time, geography, and culture. A situation-regarding comportment does not promise to yield up these sorts of generalizable truth claims . . . appreciation for the situation must become the central point of a policy discourse that anticipates action within that situation.

Such a shift to include alternative approaches in public budgeting requires an awareness of context, acceptance of a dynamic rather than static environment, and the possibility that there are multiple and valid views and intents regarding specific situations (Guba & Lincoln, 1994). When the situation in which budgeting occurs becomes increasingly diverse and uncertain (and system stability and consensus are questionable) the fit worsens with the traditional positivistic paradigm. John Morrisette (1994) likened the reliance on the technical-rational in budgeting to working with equipment that doesn't suit the environmental circumstance: he viewed budget theory and its analytical tools as approaching budgeting with a fixed wing aircraft when the conditions call for the flexibility of a helicopter where the pilot has a wider range of options with which to deal with the specifics of the situation. Approaches from other paradigms may be a better fit with budgeting realities, especially in times of "budget subtraction" (Miller, 1989, 100; Forrester & Adams, 1997).

Moving away from the approach of neutral competency requires more than technical, analytical skills – it involves understanding and often constructing and respecting a reality that is negotiated with agency stakeholders, and identifying structural barriers to inclusion. Organizational context assumes an important role, moving budgeting as a political and policy act rather than the technical neutrality forced upon it by economic paradigm (Forrester & Adams, 1997; Golembiewski, 1964). Stepping aside from the assumptions required for technical analysis, public budgeting occurs in a social world, a world that is increasingly more complex, context-dependent, interconnected and changing (Bailey & Mayer, 1992). This world in which public budgets are created is based in the interpretations, actions and interactions of participants in particular roles, each participant with responsibilities and histories related to both personal and professional agendas in political and social contexts. The case example of central support chargebacks in an Ohio state agency that was outlined at the beginning of the chapter shows the importance of including consideration of socially-constructed "realities" in addition to the accounting-based "reality" in the study of public budgeting and financial management.

PAST EFFORTS TO DEVELOP ALTERNATIVE APPROACHES TO BUDGETING

Gerald J. Miller and Alternative Budget Theories

In *Government Financial Management Theory*, Gerald Miller (1991) made a thorough but largely ignored attempt to advocate expanding legitimacy in budget theory to include what he termed alternative budget theory. His was a comprehensive attempt at deconstructing budget theories by taking a critical look at the underlying structures and assumptions of these theories. In introducing interpretive inquiry to the area of public budgeting and finance, Miller (1991) characterized existing budget theory as orthodox, prevailing, and alternative in form. These forms are detailed in Table 2.

In this conceptualization of budget theory, orthodox and prevailing approaches are based in the assumptions of system stability. Orthodox theories

Table 2. Comparison of Three Theories of Choice.

Theory	Technology	Role Theory	Organizational Process Model	Decision
Orthodox	Corporate finance	Business model	Closed system	Rational
Prevailing	Strategic management	Interest group model	Coping with uncertainty	Disjointed incrementalism
Alternative	Process design Rule design	Symbolism Myths Language	Loosely coupled system	Garbage can Retrospective rationality

Source: Miller, G. J. (1991). *Government Financial Management Theory* (p. 34). New York, Marcel Dekker, Inc.

are typified by the impartial analyst-approach to budgeting, while prevailing theories seek modifications of this impartial approach due to differences between the prescribed approach and the descriptive accounts of budget processes in the United States. According to Miller, Lindbloom's (1959) descriptive account of incrementalism typifies prevailing theories. On the other hand, Miller's alternative budget theory category has as its foundation the belief that

> . . . views or "interpretations of reality" build and gain legitimacy through an interaction of individuals. Moreover, the existence of interpretations belies the notion that there exists an objective reality shared by all organizations . . . [and,] the greater the number of different, constructed realities, the greater the uncertainty that exists among and within organizations (Miller, 1991, 7).

Working from a sociology/organizational theory perspective rather than from a purely economic or political framework, in his book Miller extended interpretive and social constructionist perspectives into public budgeting.[20] Citations from interpretive organizational approaches and philosophy of science.[21] Miller drew attention in a comprehensive fashion to the dominant metaphor of budgeting (accounting, technical assessment) and proposed an alternative approach to budgeting based upon interpretive approaches to understanding public budgeting through the lense of organizational roles created by individuals within government structures and in a public context. He suggested that the old model based in designing rational procedures to insure accountability, and spending built upon technically-based decisions no longer was adequate during times of fiscal stress and budget shortfalls – situations of ambiguity that re-emerged in public finance in the early 1980s. Miller's interpretive approach has its basis in the following premise:

> Financial management, no more or no less than any other management process, is not an ordered process deduced from some normative first principle, but a *negotiated reality constructed by the people involved* [emphasis added] . . . There is only socially constructed truth formed through intense political struggle. It follows then that these socially constructed models of financial management are unique to their own contexts, and that they emerge from the interplay of individuals (pp. 2–3).

Miller (1991, 2) suggests that socially-constructed strategies for financial viability – and therefore strategies for organizational survival, can supplant other policy-related organizational goals. Times of fiscal stress produce uncertainty that showcase political structures. This type of goal displacement was also seen in the central support chargeback example for Ohio's Department of Natural Resources that was presented earlier in the chapter. In that case, objectives relating to implementation of an acceptable chargeback system and maintaining the support of wildlife groups replaced the pursuit of other department-related policy goals.

In his characterization of orthodox theories of budgeting, Miller (1991, 79) called attention to "categorization of types of expenditures into accounts and transactions" that create an agreed-upon budget/fiscal language, that are given life through a set of procedures (e.g. year-end balances and reporting and audit requirements). Financial managers then use methods that derive from accounting principles in the creation of meaning relating to financial management and budgeting. Miller describes the language of financial management as the language of accounting, which in turn is guarded by constructed networks and systems of financial information. Accounting groupings (objects) drive the information that is kept on programs, and accounting structures are entrenched in existing budget approval systems. Accounting systems and their managers have hindered budget and other management reforms by inability/ unwillingness to keep track of program costs, and the costs of designing and implementing related accounting and record-keeping changes.[22]

Having looked at how the accounting metaphor underlies prevailing budget theory and practice, Miller next viewed executive budgeting and financial management through its rituals, signs and symbols. Here, he built upon Eldeman's (1964, 1977) emphasis on rituals in organizations and examined the roles that budgeting and accounting rituals pay in creating organizational realities. Miller (1991, 101) concluded that financial management is largely symbolic in nature:

> . . . a budget manager's function is that of manipulating symbols, producing rituals, and employing a unique language to get the budget and impose it . . . [Further] finance departments create the reality that organizations have by symbolically, ritualistically, and rhetorically coping with the most critical problem each organization has, namely resource constraints.

To support this claim, Miller used the example of budget reforms in the 1980s in the state of Kansas (Miller, Rabin & Hildreth, 1987, 108–113). In this case, the state budget director and governor created the requirement that agency budget proposals must offset increases to a base level with spending reductions. This balanced base budget system became a symbol for the changing subcultures of budget power within the state, particularly the governor's resurgence as the dominant actor in the state's budget formation and control. Certainly similar jockeying is also seen in the balanced budgets submitted by U.S. presidents, especially after the Gramm-Rudman-Hollings Act and other balanced budget amendment initiatives in Congress in the early and mid 1990's – even though the balanced budget submission is understood by many participants and observers as created by smoke and mirrors.

We can extend Miller's analysis into the 1990s. Located within White and Adam's (1995) 7th narrative, public management, the enacted metaphor of the balanced budget can be deconstructed through the realm of critical and postmodern analysis. The ambiguities and complexities of the U.S. budget were simplified to a language of offsetting revenues and new programs – no net increases to the budget. Miller (1991), and other public administration theorists applying postmodern paradigms, predict that a simplifying metaphor remains useful and orthodox if it maintains the stability of office holders, citizens do not demand changes, and agencies maintain their relative positional strengths (Beresford, 2000; Farmer, 1995; Fox & Miller, 1995).

Creating a new metaphor can be destabilizing, as seen in the next example involving systemic funding changes in Ohio's education system, necessitated by an Ohio Supreme Court decision.

School Funding in Ohio

A recent state-wide ballot issue related to primary and secondary school funding in Ohio provides an example of the consequences for political leaders who do not successfully create an acceptable metaphor for public finance matters. Like many states, Ohio counties vary in average income and property tax values (Ohio Department of Taxation Report, 1998, 1997). The majority of school districts are funded through local property taxes as well as through a state-wide formula that primarily is based on enrollment.[1] In 1991, Bill Phillis, a retired administrator from the State Board of Education formed the Coalition for Equity and Adequacy of School Funding, which filed a court suit charging inequality in the state's school district funding formula based upon disparities across the state spent on education (with rural and densely urban districts spending lower amounts than suburban school districts). The case [DeRolf vs.The State of Ohio] went to the Ohio Supreme Court, which found for the plaintiff in March 1997 that the existing school funding formula was inequitable. The Court required Ohio's political leaders (Governor and Legislature) to provide an equitable funding process by March 31, 1998.

The compromise funding strategy was a proposed sales tax increase of 1 cent per dollar (1%) to be earmarked for primary and secondary education. A small reduction in residential property taxes was included in the deal – a maximum property tax credit of $275 per year. This legislation (Am Sub HB 679) was passed by the General Assembly and signed by the Governor in February 1998, and included the provision that the 1% sales tax increase be placed on the May 1998 state-wide ballot for voter approval. Despite further legal challenges, the sales tax issue was placed on the May 1998 ballot and was resoundingly defeated by voters (80% against). Perhaps the issue was never acceptably framed for the public. The Governor promoted the tax by comparing the maximum reduction of the residential property tax, $275, with the total amount of purchases necessary to override the property tax relief, $27,500.[2] The average tax increase per family was promoted in state-wide campaign advertisements as an estimated $125 per year.

Regressiveness of the sales tax as a funding source was not featured in the promotion – in Ohio clothing is taxed, with food the only major consumable not taxed. Taxpayer reticence might be attributed to the public wanting better schools, lacking belief that more money means better schools (although Ohio ranks 22nd in primary and secondary per pupil spending nationally),[3] reluctance to pay more taxes (the Governor's testimony notwithstanding); or lack of persuasion of the value of adopting an incremental solution to the school funding program.

On May 11, 2000 the Ohio Supreme Court ruled a second time that Ohio's school funding system is unconstitutional. It is back to the drawing board for state officials and for Governor Taft on the school funding issue. This go-round, the Governor plans to "help bring more people into the debate."[4]

1. Since 1989, school districts may also utilize voted income tax support: seven districts now include income tax as a revenue source
2. Leonard (1998).
3. Fall enrollment, 1999, as reported in *Ranking of the States: 1999 and 2000 Estimates of School Statistics*. National Education Administration, 1999.
4. See Alan Johnson and Lee Leonard, "Taft plans to travel state to gather ideas on school funding," Columbus Dispatch, June 13, 2000. See also http://www.OhioSchoolFunding.org, the Governor's school funding web site.

A deconstructive and critical analysis of this approach would incorporate consideration of the solutions that were not offered, the language surrounding the framing and selling of the issue to the public, and consideration of who had participation access to the solution, as well as the barriers to participation. The social welfare issue of redistribution of wealth that underlies decades of school funding differentials between Ohio's suburban districts and rural and urban districts went undiscussed in issue promotions, as did the issue of local control of schools. Given the short time period allowed by the Ohio Supreme Court, public debate centered only on technical solutions.

From his interpretive, critical framework, Miller (1991, vi) described that in creating meaning surrounding budget concerns and issues,

> The battle waged in financial management, and in organizations that have an important concern with financial matters, is one over the interpretation of complex events. Financial managers find themselves engaged in interpreting, and finally gaining the upper hand in determining critical assumptions. The value of values is clearly the determination to be made in valuing budgets and parts of budgets, in short. Derived processes – interpretation through symbol and metaphor, for example – are means of indirection in determining and gaining consensus over the value of values.

Similarly, Farmer (1996, 132) concluded (from the philosophy of Foucault) that "[w]hat is counted as real knowledge is determined by interacting sets of strategic power considerations and relationships." In the case of Ohio's school funding problem, the established political power structure collectively created meaning around a funding "solution" that could not be sold to voters in the associated state-wide ballot issue.

FURTHERING A POSTMODERN NARRATIVE IN PUBLIC BUDGETING AND FINANCE

Miller's examination of the underlying assumptions and the consequences of the prevailing approaches to budgeting, and his acknowledgment of the entrenched interests in budgeting, also led to advocating the idea of budgeting that has as its basis collective action. According to Miller (1991, 7), "Orthodox and prevailing theory depend for their explanatory power on relative large amounts of consensus on organizational goals and technologies . . . Alternative theory also seeks the fundamental, intersubjectively determined premises that make *collective action* possible." (emphasis added). And, alternative theory does not assume certainty or stability (p. 8).

Collective action becomes the groundwork for applied critical and action approaches to public budgeting and financial management. In order for collective action to influence budget policy, the existing barriers to legitimate participation by those excluded from the dominant dialogue must be identified and removed. The emergent paradigm in public budgeting relies upon local narratives and collective action based in all voices.

Removing Structural Barriers

Public administrators have used various control methods to keep citizens out of decision processes, including limiting citizen input until a time when it is meaningless or an unessential part of the policy process (King & Stivers, 1998). This control is especially apparent in the budget process. Typically budgets are developed internally by the executive branch and are presented to the public and the legislature as a finished product, or one that can be tampered with only marginally. This approach relates to the executive budget process established in the 1921 Budget and Accounting Act, which resulted in the transfer of Constitutional powers from the Congress to the President (Bailey, 1989). In concert with the emphasis of the political reform movement on business methods, professional expertise was given more value than democracy. Too much participation created inefficiencies.[23]

Public administration theory is again cycling towards valuing participation – this time "authentic" participation (King & Stivers, 1998). Managers have discovered that working collaboratively with citizens leads to a better allocation of resources (Gray & Chapin, 1998). In effect, operating from a critical value structure that places citizen views as equal allows administrative experts to step away from the confines and limitations of budgeting as accounting to hear the values of the community and include or base the budget on these values (Rubin, 1996). Essentially this produces local budget narratives and choices detailed to each local context.

For example, Gray and Chapin (1998) reported on a project involving budget collaboration that produced outcomes outside the usual forms produced from following the traditional accounting framework. Labeled the "Targeted Community Initiative," Orange County, Florida administrators sought community involvement over a three year period. After working with one community, they realized they were too heavy handed in their control and direction, and revised their participatory mechanisms. In the end, administrators became facilitators for the achievement of community-identified priorities. And, they found that involving the citizens in discussing community initiatives led to different mechanisms for service delivery that were not all provided by government. Community participation removed the assumption of administrative expertise in policy and budget issues. This process took longer than producing administratively-set priorities, but Gray and Chapin viewed the outcome as preferred from the standpoint of allocating limited resources. The pool of resources and means of program achievement moved from government-only to include other sectors and resources at the initiative of citizens. postmodern budgeting will be focused on local narratives, allowing adjustments and contextual interpretations and actions.

THE POSTMODERN CONNECTION TO INTERPRETIVE AND CRITICAL BUDGETING: IMPLICATIONS FOR DOING RESEARCH

Claiming that an understanding of context and the socially-constructed aspects of budgeting are necessary for developing alternative budget theory requires adjustments in the way that we conduct and present our research. Without awareness of and adherence to research strategies that respect related ontological differences, researchers/budget theorists cannot claim that their work involves an interpretive or critical approach to budgeting (although few to date have wanted to).[24] These differences in approaches are important because not only are there differences in the types of knowledge about budgeting – the academic study of budgeting, but the process of DOING budget theory – in these alternative paradigms is different, too.

Interpretive research is *inductive*, focused on the sensemaking and constructions of participants rather than on the tacit and formal theories of the researcher. Marshall and Rossman (1995) distinguished between tacit theory as the researcher's theory-in-use and formal theory as that obtained from literature reviews and existing social science theory. Each influences the researcher's perspective, pushing research into a deductive mode and influencing qualitative research processes. Writing about public administration research, White and Adams (1994, 11) stated that the "technical rationality view [is so ingrained that] . . . the scientific-analytic mind-set has to be almost `unlearned' before other, alternative ways of knowing can be grasped." Applying these appraisals to budget research, descriptive and deductive approaches to research result in what Miller (1991) termed prevailing and orthodox budget theories.

Research that is based in the views of the lived experiences of budget participants (see Rubin, 1998; Rubin & Rubin, 1995 as examples) requires naturalistic or field-based inquiry strategies (Denzin, 1989; Denzin & Lincoln, 1994; Lincoln & Guba, 1985). Approaching research with an openness to a subjective world of participants rather than a world of objective and verifiable truths guides many research decisions.[25] Interpretive and critical ontologic assumptions are reflected in framing the research question, in research design issues such as sampling, question format, and data analysis processes, as well as in the written research presentation. I have written elsewhere about these related research decisions in the context of research in public administration (Orosz, 1997, 1998). The assumptions of naturalistic inquiry and how they relate to interpretive, critical, and postmodern forms of budgeting and budget analysis, as well as how one conducts interpretive inquiry in the area of public budgeting are topics of sufficient scope to be addressed separately.[26] Elaboration of these ideas and how they apply to those who wish to conduct budget research from other paradigms are left for future work.

SO, IS POSTMODERN BUDGETING A REAL DEAL?

In the first part of this chapter, I addressed the potential value of adding interpretive and critical approaches to the study of public budgeting. Here, I return to comments on the more general topic from public administration theorists for assessments of the usefulness of alternative approaches to knowledge (getting to the question if this postmodern budgeting is a "real deal"). These theorists addressed the question "What is required for alternative approaches [to knowledge development] to contribute to the development of knowledge in public administration theory?" White's (1986a, 232) opinion is that "Non-mainstream research can contribute to knowledge in public administration *if it is guided by alternative methodological and philosophical frameworks*" (emphasis added). Research conducted with the idea of identifying interpretive realities of participants in various roles, concerning a specific social situation, too often still seeks nomothetic generalization.[27] A key challenge for alternative paradigm researchers is relinquishing the goal of making general statements about the topic studied, and eliminating associated language from written presentation of the material. The difficulties in relinquishing these research conclusions hinge on concern about whether and how, in the absence of general conclusions, if these alternative modes of inquiry can produce research that is useful to the advancement of public administration.

One standpoint from which to consider critical theory/deconstructionist approaches is the related academic system and what it produces or collectively adopts as knowledge (Farmer, 1996, 1997; Fox & Miller, 1995, 1997; Richardson, 1997; White & Adams, 1994; White, 1999). This also raises the question of the privileged role of the researcher and the consequences of the structure of scientific-style inquiry to the practice and study of budgeting. The decade in which Miller's (1991) book was overlooked leads me to suggest that the "conversational community" of public budgeting and finance is only now opening to these alternative forms of knowledge.[28]

Summary and Future Directions

In this chapter, I advocated the possibilities for including alternative (interpretive, critical, and postmodern) approaches to the study and conduct of public budgeting. In so doing, I discussed narratives of public administration and how these do or do not transfer to the study of public budgeting, and the usefulness of these different approaches to the study of public budgeting. Gerald Miller's (1991) contribution to the development of alternative approaches to budget theory was reviewed, and I discussed some of the structural and cultural impediments to the advancing of these types of alternative theories. Examples based in budgeting practice and public administration theory were given to show the operational and theoretical consequences of limiting budget and policy actions to those in the technical/rational paradigm.

I hope to have rekindled interest in working from alternative paradigms in the practice and study of public budgeting and finance. The fact that this chapter was included in this book on developments of budget theory is evidence that there can be room in the dialogue of budget theorists for consideration of postmodern and other alternative approaches to budgeting. The next step is to develop frameworks and approaches that allow broad participation, interpretations, critical analysis and deconstruction, communication, and discourse. The experiences of budget practitioners reveal that the truth*s* are being created out there. We need only come to know it and realize its positioning, values, and constraints.

NOTES

1. See also Woller, Gary M. (Ed.) (1997). Administration and Postmodernism (special topic issue). *American Behavioral Scientist*, *41*(1).

2. Poststructualists "focus on the interpretation of text and symbol systems" (White, 1999, 163). White also provides an abbreviated summary of the contributions of Jacques Lacan (1977, 1978) and Jacques Derrida (1973, 1976, 1978) to post-structuralism and the study of public administration, as do Farmer (1995) and McSwite (1977).

3. Other chapters in this book more completely review specific aspects of public budget theory as presently constructed.

4. My apologies to Fred Thompson for any oversimplification/misrepresentation of his remarks.

5. Each of these authors describes and comments on the utility of critical and postmodern approaches to the study of public administration.

6. As does modernism have its own language.

7. The study of budgeting from a "political science" orientation on politics seems closest to this approach (see Wildavsky,1992; and Rubin, 1997 for examples of leading texts).

8. Others are not as generous in their assessments of the progress in the direction of public administration's movement toward postmodernism/ critical, interpretive direction.

See Dennard (1997), Orosz (1997), Zanneti (1998), Marshall and Choudhury (1997) for these assessments and related prescriptions.

9. See White and Adams (1994) for examples of this position regarding public administration in general.

10. Described by Timney (1995, 44–45).

11. This praxis, or interrelationship of theory and practice, narrative continues in arguments considering accreditation programs for finance and budget offices (Hildreth, 1998).

12. Frank Marini (1992) commented that the shift toward the New Public Administration, as a paradigm, did not reach its potential.

13. For example, see the National Association of Schools of Public Affairs and Administration's (NASPAA) web page on requirements for Accounting and Finance professionals:

> The local government administrator needs to go far beyond budgeting. Familiarity with accounting and financial reporting, the assessment of financial conditions, knowledge of creative financing techniques, capital financing methods, and cash management are essential. . . . Because they are involved in revenue as well as expenditure policy development, local administrators must understand basic principles of public finance and tax policy.

14. See for example Hummel (1977); (Harmon, 1981).

15. As another example of the juxtaposition of politics with economic rationality for budget analyst decision making, Thurmaier (1992) proposed and experimentally tested a model of the multiple rationalities used by budgeteers in making funding recommendations. He concluded that availability and timing of budgetary and political information impacted the recommendations of the budget analyst. When students in the sample were compared with practicing budgeteers, the students were more likely to make recommendations based on technical criteria if the political criteria were not provided beforehand (budgeteers with experience at better at anticipating the political information). Applying this to the case of central support chargebacks used earlier in this chapter, the focus on the financial problems of the state of Ohio and the need to locate additional resources preceded the show of strength of the Division of Wildlife political groups – reflecting the tensions of the political context and the economic and technical chargeback concerns.

16. Specifically, Liggett used narrative from *The Milagro Beanfield War* (1974) by John Nichols in her literary analysis of CDBG (New York: Holt, Rinehart and Winston).

17. cited in Timney 1995, p 47.

18. Informal survey of budget course syllabi. Of course, this is a general statement, and some courses are "politics-of-budgeting" in their orientation rather than economics and technical-skill-based.

19. Miller's suggestions were of a cautious nature, and did not attack the legitimacy of rationalist approaches in offering a look at alternative ways of considering budgeting and financial management. Still, the prescription was not acted upon and received limited attention through the 1990s (even by Miller, who pursued other topics in public finance). Alexander (1999) and Kearns (1995) have each proposed alternative models; Alexander dealing with the need for a democratic ethos in budgeting, and Kearns focusing on broader interpretations of budget accountability. Each proposal is based upon discussions with budgeters about their experiences.

20. In the 1980s renewed attention was focused on including and extending interpretive forms of knowledge acquisition to the study of organizations

21. Constructionist approaches built upon the work of Berger and Luckman's (1966) *Social Construction of Knowledge*, Burrell and Morgan's (1979) *Organisational Analysis* and Weick's (1979) *The Social Psychology of Organizing*, then moved into the construction of organizational cultures and metaphors of organizations (Schein, 1985; Morgan, 1986).

22. However, proposals to reforms these systems (Gore, 1993) generally maintain the existing prevailing managerial and political system. See Patricia W. Ingraham, James R. Thompson, and Ronald P. Sanders, 1997. *Transforming Government: Lessons from the Reinvention Laboratories*. San Francisco: Jossey-Bass.

23. The following are some of the cities that have or had exemplary citizen participation in budget preparation and decision processes: Dayton, Ohio; Eugene, Oregon; and Seattle, Washington.

24. However, throughout her career, Irene Rubin has conducted much of her budget research through formal interviews with high-ranking budget officials (Rubin & Rubin, 1995).

25. This interview approach, focused on budget participants, is itself entrenched in the existing definitions of public budgeting and its existing participants. A critical analysis focusing on who is and is not participating, and other exclusions would start to move beyond prevailing and orthodox descriptions.

26. See Yanow (2000).

27. 34. See Fox and Miller (1995).

28. My opinion, not Gerald Miller's.

ACKNOWLEDGMENTS

Many thanks to Mary Timney for her assistance in the early development of this chapter, and to Gerald Miller for his suggestions for the manuscript and his guidance for future directions in the development of critical approaches to budgeting. I would also like to acknowledge this Volume's editor, John Bartle, for his patience, support, and gentle prodding that led to the completion of this chapter (only occasionally was it like a cattle prod).

REFERENCES

Adams, G. B. (1992). Enthralled with Modernity: The Historical Context of Knowledge and Theory Development in Public Administration. *Public Administration Review*, *52*(3), 363–373.

Alexander, J. (1999). A New Ethics of the Budgetary Process. *Administration and Society*, *31*(4), 542–565.

Ammons, D. N. (1991). *Administrative Analysis for Local Government: Practical Applications of Selected Techniques*. Athens, GA: University of Georgia Carl Vinson Institute of Government.

Axelrod, D. (1995). *Budgeting for Modern Government* (2nd ed.). New York: St. Martin's Press.

Bailey, M. T. (1989). Budgeting Theory in the 1980s: Alive and Not So Well. *New Directions in Public Administration Research*, *2*(1), 85–95.

Bailey, M. T., & Mayer, R. T. (Eds) (1992). *Public Administration in an Interconnected World: Essays in the Minnowbrook Tradition*. Westport, CT: Greenwood Press.

Berger, P., & Luckmann, T. (1966). *The Social Construction of Reality: A Treatise in the Sociology of Knowledge*. Garden City, NY: Doubleday.

Beresford, A. D. (2000). Simulated Budgeting. *Administrative Theory and Praxis, 22*(3), 479–497.

Burrell, G., & Morgan, G. (1979). *Sociological Paradigms and Organizational Analysis*. Portsmouth, N.H.: Heinemann.

Cope, G. H. (1989). Yes, Professor Key, There Is a Budget Theory: Three Premises for Public Budgeting. *New Directions in Public Administration Research, 2*(1), 75–82.

Chock, P. P. (1995). Ambiguity in Policy Discourse: Congressional Talk about Immigration. *Policy Sciences, 28*(2), 165–184.

Colebatch, H. K. (1995). Organizational Meanings of Program Evaluation. *Policy Sciences, 28*(2), 149–164.

Denhardt, R. B. (1984). *Theories of Public Organization*. Monterey, CA: Brooks/Cole.

Dennard, L. F. (1997). The Democratic Potential in the Transition of Postmodernism: From Critique to Social Evolution. *American Behavioral Scientist, 41*(1), 148–162.

Denzin, N. K. (1989). *Interpretive Interactionism*. Newbury Park, CA: Sage.

Denzin, N. K., & Lincoln, Y. S. (Eds) (1994). *Handbook of Qualitative Research*. Newbury Park, CA: Sage.

Derrida, Jacques. (1973). *Speech and Phenomena, and Other Essays on Husserl's Theory of Signs*. Evanston, IL: Northwestern University Press.

Derrida, J. (1976). *Of Grammatology*. Baltimore MD: Johns Hopkins University Press.

Derrida, J. (1978). *Writing and Difference*. London: Routledge and Kegan Paul.

Eldleman, M. (1977). *Political Language: Words That Succeed and Policies That Fail*. New York: Academic Press.

Eldleman, M. (1964). *The Symbolic Uses of Politics*. Urbana, IL: University of Illinois Press.

Farmer, D. J. (1997). Derrida, Deconstruction, and Public Administration. *American Behavioral Scientist, 41*(1), 12–27.

Farmer, D. J. (1996). The postmodern Turn and the Socratic Gadfly. *Administrative Theory and Praxis, 18*(1), 128–133.

Farmer, D. J. (1995). *The Language of Public Administration*. Tuscaloosa: University of Alabama Press.

Fischer, F. (1981). *Politics, Values, and Public Policy: The Problem of Methodology*. New York: Westview Press.

Fischer, F., & Forester, J. (Eds) (1993). *The Argumentative Turn in Policy Analysis and Planning*. London: Duke University Press.

Forrester, J. P., & Adams, G. B. (1997). Budgetary Reform Through Organizational Learning: Toward an Organizational Theory of Budgeting. *Administration and Society, 28*(4), 466–488.

Fox, C. J., & Miller, H. T. (1995). *Postmodern Public Administration: Towards Discourse*. Thousand Oaks, CA: Sage.

Fox, C. J., & Miller, H. T. (1997). The Depreciating Public Policy Discourse. *American Behavioral Scientist, 41*(1), 64–89.

Golembiewski, R. T. (1964). Accountancy as a Function of Organization Theory. *Accounting Review, 39*(April), 333–341.

Gore, A. G. (1993). *From Red Tape to Results*. Washington, D.C.: U.S. Government Printing Office.

Gosling, J. S. (1987). State Budget Office and Policy Making. *Public Budgeting & Finance, 7*(Spring), 51–65.

Gray, J. E., & Chapin, L. W. (1998). Targeted Community Initiative: Putting Citizens First! In: C. Simrell King, C. Stivers & Collaborators (Eds), *Government Is Us: Public Administration in an Anti-government Era* (pp. 175–194). Thousand Oaks, CA: Sage.

Griffith, D. M., & Associates (1984). *A Central Services Cost Allocation Plan (Draft) Ohio Department of Natural Resources, Budgeted Costs for the Year Ended June 30, 1985.* Columbus, OH.

Guba, E. G., & Lincoln, Y. S. (1994). Competing Paradigms in Qualitative Inquiry. In: N. Denzin & Y. S. Lincoln (Eds), *Handbook of Qualitative Research* (pp. 105–117). Thousand Oaks, CA: Sage.

Gulick, L., &. Urwick, L. (Eds) (1937). *Papers on the Science of Administration.* New York: Institute of Public Administration.

Harmon, M. M. (1981). *Action Theory for Public Administration.* New York: Longman.

Harmon, M. M. (1995). *Responsibility as Paradox: a Critique of Rational Discourse in Government.* Thousand Oaks, CA: Sage.

Hildreth, W. B. (1998). Should There Be an Accreditation Program for Finance and Budget Offices? *Public Budgeting & and Finance, 18*(2), 18–23.

Hummel, R. P. (1977). *The Bureaucratic Experience* (1st ed.). New York: St. Martin's Press.

Ingraham, P., Thompson, J. R., & Sanders, R. P. (Eds) (1997). *Transforming Government: Lessons from the Reinvention Laboratories.* San Francisco: Jossey-Bass.

Jun, J. S., & Rivera, M. A. (1997). The Paradox of Transforming Public Administration: Modernity Versus Postmodernity Arguments. *American Behavioral Scientist, 41*(1), 132–147.

Kelly, M., & Maynard-Moody, S. (1993). Policy Analysis in the Post-positivist Era: Engaging Stakeholders in Evaluating the Economic Development Districts Program. *Public Administration Review, 53*(2), 135–142.

Kearns, K. P. (1995). Accountability and the Entrepreneurial Public Manager: The Case of the Orange County Investment Fund. *Public Budgeting & Finance, 15*(3), 3–21.

Key. V. O. (1940). The Lack of a Budgetary Theory. *American Political Science Review, 34*(6), 1137–1144.

King, C. S., Stivers, C., & Collaborators (1998). *Government Is Us: Public Administration in an Anti-government Era.* Thousand Oaks, CA: Sage.

Lacan, J. (1977). *Ecrits: A Selection.* London: Travistock.

Lacan, J. (1978). *The Language of the Self – the Function of Language in Psychoanalysis.* Baltimore MD: Johns Hopkins University Press.

Lauth, T. P. (1978). Zero-base Budgeting in Georgia State Government: Myth and Reality. *Public Administration Review, 38*(5), 420–430.

Leonard, L. (April 24, 1998). Voinovich Hits Home with Issue 2 Campaign Stop. Columbus Dispatch, p. 4B.

Liggett, H. (1997). Play of Approved Method and Authorized Value in the Administration of Public Issues. *International Journal of Public Administration, 20*(11), 1955–1978.

Lindblom, C. E. (1959). The Science of Muddling Through. *Public Administration Review, 29*(spring), 79–88.

Lincoln, Y. S. (1995). Emerging Criteria for Quality in Qualitative and Interpretive Research. *Qualitative Inquiry, 1*(3), 275–289.

Lincoln, Y. S., & Guba, E. G. (1985). *Naturalistic inquiry.* Beverly Hills, CA: Sage.

Linder, S. K. (1995). Contending Discourses in the Electric and Magnetic Fields Controversy: the Social Construction of EMF Risk as a Public Problem. *Policy Sciences, 28*(2), 209.

Lynn, L. E. (1996). *Public Management as Art, Science, and Profession.* Chatham, N.J.: Chatham House Publishers.

McSwite, O. C. (1997). Jacques Lacan and the Theory of the Human Subject: How Psychoanalysis Can Help Public Administration. *American Behavioral Scientist, 41*(1), 43–63.

Marini, F. (1992). Introduction. In: M. Bailey & R. T. Mayer (Eds), *Public Administration in an Interconnected World.* Westport, CT: Greenwood.

Marini, F. (Ed.) (1971). *Toward a New Public Administration: The Minnowbrook Tradition*. Scranton, PA: Chandler.

Marshall, C., & Rossman, G. (1995). *Designing Qualitative Research* (2nd ed.). Thousand Oaks, CA: Sage.

Marshall, G. S. & Choudhury, E. (1997). Public Administration and the Public Interest. *American Behavioral Scientist*, *41*(1), 119–131.

Mikesell, J. L. (1995). *Fiscal Administration: Analysis and Applications for the Public Sector* (3rd ed.). Belsmont, CA: Wadsworth Publishing Co.

Miller, G. J. (1991). *Government Financial Management Theory: Public Administration and Public Policy*. In: J. Rabin (Ed.). *Series on Public Administration and Public Policy* (Vol. 43). New York: Marcel Dekker, Inc.

Miller, G. J. (1989). V.O. Key and the Ambiguities of Choice. *New Directions in Public Administration Research*, *2*(1), 97–101.

Miller, G. J., Rabin, J., & Hildreth, W. B. (1987). Strategy, Values, and Productivity. *Public Productivity Review*, *11*(1), 81–96.

Morrisette, J. R. (1994). Rationality in Public Budgeting: Fission in a Fusion World. Unpublished Ph.D. Dissertation, University of Cincinnati.

Morgan, G. (1986). *Images of Organization*. Newbury Park, CA: Sage.

NASPAA. (1992). Guidelines for Local Government Management Education (ICMA/NASPAA Task Force on Local Government Education). Retrieved from the World Wide Web, 12/28/99. Http://www.naspaa.org/guidelines/guide_logoved.html

Napier, W. J. (1989). Interview with J. Foley Orosz. November 6.

National Education Administration. (1999). *Ranking of the States: 1999 and 2000 Estimates of School Statistics*, p. 55.

Nichols, J. (1974). *The Milagro Beanfield War*. New York: Random House.

Orosz, J. F. (1997). Resources for Qualitative Research: Advancing the Application of Alternative Methodologies in Public Administration. *Public Administration Review*, *57*(6), 543–549.

Orosz, J. F. (1998). Widening the Yellow Brick Road: Answering the Call for Improved and Relevant Research in Public Administration. In: J. D. White (Ed.), *Research in Public Administration* (Vol. 4, pp. 87–104). Greenwich, CT: JAI Press.

Pal, L. A. (1995). Competing Paradigms in Policy Discourse: The Case of International Human Rights. *Policy Sciences*, *28*(2), 185–208.

Pyhrr, P. A. (1973). *Zero-base Budgeting: A Practical Management Tool for Evaluating Expenses*. New York: John Wiley.

Richardson, L. (1997). *Fields of Play: Constructing an Academic Life*. New Brunswick, NJ: Rutgers University Press.

Rubin, I. S. (1998). *Class Tax and Power: Municipal Budgeting in the United States*. Chatham, NJ: Chatham House Press.

Rubin, I. S. (1997). *The Politics of Public Budgeting: Getting and Spending, Borrowing and Balancing* (3rd ed.). Chatham, NJ: Chatham House Publishers.

Rubin, I. S. (1996). Budgeting for Accountability: Municipal Budgeting for the 1990s. *Public Budgeting & Finance*, *16*(2), 112–132.

Rubin, I. S. (1994). Early Budget Reformers: Democracy, Efficiency, and Budget Reforms. *American Review of Public Administration*, *24*(3), 229–251.

Rubin, I .S. (1993). Who Invented Budgeting in the United States? *Public Administration Review*, *53*(5), 438–444.

Rubin, I. S. (1991). Budgeting for Our Times: Target Based Budgeting. *Public Budgeting & Finance*, *11*(3), 6–19.

Rubin, I. S. (1990). Budget Theory and Budget Practice: How Good the Fit? *Public Administration Review, 50*(2), 179–189.

Rubin, H. J., & Rubin, I. S. (1995). *Qualitative Interviewing: The Art of Hearing Data*. Thousand Oaks, CA: Sage.

Schein, E. H. (1985). *Organizational Culture and Leadership: A Dynamic View*. San Francisco: Jossey-Bass.

Schick, A. (1966). The Road to PPB: The Stages of Budgetary Reform. *Public Administration Review, 26*(6), 234–258.

Schick, A. (1978). The Road From ZBB. *Public Administration Review, 38*(2), 177–180.

State of Ohio, Department of Taxation (1997). School District Taxable Property Values: Taxes Levied and Tax Rates for Current Expenses, and Average Property Values per Pupil, Tax Year 1996. Table SD-1, no. 55, July 7.

State of Ohio, Department of Taxation (1998). Personal Income Tax: 1996 Income Tax Returns by School District. Table Y-2, No. 52.

Stivers, C. (1993). *Gender Images in Public Administration*. Newbury Park, CA: Sage.

Thurmaier, K. (1992). Budgetary Decisionmaking in Central Budget Bureaus: An Experiment. *Journal of Public Administration Research and Theory, 2*(4), 463–487.

Thurmaier, K. (1995). Decisive Decision Making in the Executive Budget Process: Analyzing the Political and Economic Propensities of Central Budget Bureau Analysts. *Public Administration Review, 55*(5), 448–460.

Thurmaier, K., & Gosling, J. J. (1997). The Shifting Roles of State Budget Offices in the Midwest: Gosling Revisited. *Public Budgeting & Finance, 17*(4), 48–70.

Timney, M. M. (1995). The Asynchronism of Public Administration and Budget Theories. *Administrative Theory and Praxis, 17*(2), 41–51.

Weick, K. E. (1979). *The Social Psychology of Organizing* (2nd ed.). New York: McGraw-Hill.

White, J. D. (1999). *Taking Language Seriously: The Narrative Foundations of Public Administration Research*. Washington, D.C.: Georgetown University Press.

White, J. D. (1992). Taking Language Seriously: Toward a Narrative Theory of Knowledge for Administrative Research. *American Review of Public Administration, 22*(2), 75–88.

White, J. D. (1986a). Dissertations and Publications in Public Administration. *Public Administration Review, 46*(3), 227–234.

White, J. D. (1986b). On the Growth of Knowledge in Public Administration. *Public Administration Review, 46*(1), 15–24.

White, J. D., & Adams, G. B. (1995). Reason and Postmodernity: the Historical and Social Context of Public Administration Research and Theory. *Administrative Theory & Praxis, 17*(1), 1–19.

White, J. D., & Adams, G. B. (Eds) (1994). *Research in Public Administration*. Newbury Park, CA: Sage.

Willoughby, K. G. (1993). Patterns of Behavior: Factors Influencing the Spending Judgements of Public Budgeteers. In: T. Lynch & L. Martin (Eds), *The Handbook of Comparative Public Budgeting and Financial Management*. New York: Marcel Dekker.

Wildavsky, A. (1964). *The Politics of the Budgetary Process*. Boston: Little, Brown and Co.

Wildavsky, A. (1992). *The New Politics of the Budgetary Process* (2nd ed.). New York: Harper Collins.

Wildavsky, A. (1986). *Budgeting: a Comparative Theory of Budget Processes* (2nd rev. ed.). Piscataway, N.J.: Transaction Books.

Wilson, W. (1887). The Study of Administration. *Political Science Quarterly, 66*(December 1941, reprinting original), 481–506.

Woller, G. M. (Ed.) (1997). Public Administration and Postmodernism (special topic issue). *American Behavioral Scientist, 41*(1).

Woller, G. M., & Patterson, K. D. (1997). Public Administration Ethics. *American Behavioral Scientist, 41*(1), 103–118.

Yanow, D. (2000). *Conducting Interpretive Policy Analysis*. Thousand Oaks: Sage.

Yanow, D. (1996). *How Does a Policy Mean: Interpreting Policy and Organizational Actions*. Washington D.C.: Georgetown University Press.

Zanetti, L A. (1998). At the Nexus of State and Civil Society: The Transformative Practice of Public Administration. In: C. S. King, C. Stivers & Collaborators (Eds), *Government Is Us: Public Administration in an Anti-government Era*. Thousand Oaks, CA: Sage.

Zanetti L. A. (1997). Advancing Praxis: Connecting Critical Theory With Practice in Public Administration. *American Review of Public Administration, 27*(2), 145–167.

8. APPLYING TRANSACTION COST THEORY TO PUBLIC BUDGETING AND FINANCE

John R. Bartle and Jun Ma

ABSTRACT

This paper applies transaction cost theory to public budgeting and finance. First, related literature is examined, focusing on Horn's transaction cost model of public administration, Dixit's transaction cost politics approach, the transaction cost budget model of Patashnik, and Thompson and Jones' model of budgetary execution. Second, it is argued that transaction cost politics rather than transaction cost economics is more suitable for understanding the public sector in general and public budgeting in particular. It is also argued that the focus on minimizing transaction cost is problematic. Third, this paper discusses some building blocks for a transaction cost theory of public budgeting. Fourth, potential applications of this theory to subjects in public budgeting and financial management are discussed. It is concluded that although transaction cost theory has some shortcomings when applied to the study of public budgeting, it is a promising and attractive approach.

Evolving Theories of Public Budgeting, Volume 6, pages 157–181.

ISBN: 0-7623-0790-0

INTRODUCTION

Transaction cost theory has recently had an important influence on economics. It would seem naturally suited for application to the study of public budgeting and finance for several reasons. First, its concepts can be applied to budgeting and finance fairly easily. Budget agreements and deals are transactions of a sort, and opportunism, uncertainty, and information asymmetry abound in budgetary politics. Second, transaction cost theory is centrally focused on institutions and history, unlike many other economic models. In an inherently institutionally-based field of study such as public administration, this makes for a comfortable marriage. Finally, the generality of transaction cost theory could serve as a vehicle to organize the disparate findings and proverbs of public budgeting into a more coherent conceptual framework. This would allow for better comparative research, and for bringing a historical perspective to theoretical explanations.

Despite the natural connection, there has been little progress in the application of transaction cost theory to public budgeting and finance. This paper reviews the relevant literature to date, and then identifies the basic building blocks of such a theory. While it is far too premature to pass any judgment on the utility of such a theory, we do believe that it has promise, although certain issues loom large.

LITERATURE REVIEW

Transaction cost theory was developed within economics and focuses on explaining organizational form in the private sector. It has also been used to address issues in public administration, politics, budgeting institutions, budgetary execution, and fiscal policy. Here, we examine in detail some of the most significant works in these areas: Horn's transaction cost model of public administration, Dixit's transaction cost politics approach, the transaction cost budget model of Patashnik, and Thompson and Jones' model of budgetary execution. Other important works not discussed in detail here include: Butler (1983), Moe (1984, 1989, 1990a, b), Maser (1986), Weingast (1984), Bryson and Ring (1990), Heckathorn and Maser (1987), Vining and Weimer (1990), Weimer (1992), Twight (1994), and Pouder (1996).

A Transaction Cost Model of Public Administration

Horn (1995) has developed a transaction cost model of public administration, which focuses on examining the effects of transaction problems legislators face on their choice of institutions. To Horn, modern governmental institutions, including budgetary institutions, serve the legislature's interests well through solving the transaction problem. Horn's framework attempts to explain institutional characteristics of the administrative branch of government. His model focuses on the institutional choices of the legislature. He starts with a situation in which issue complexity and the desire to take advantage of specialization lead representative assemblies to rely on bureaucratic agents. Since the exchange between legislators and their constituents is typically not simultaneous, he argues that legislators creating agency missions need to add durability to their deals by protecting the agency against subsequent legislative modification, while protecting themselves against agency actions detrimental to their interests. He argues that institutional structures in the public sector are designed to serve the interests of the legislators who create them. Therefore, the nature of these institutional structures is determined by the transaction problems the legislators meet.

He indicates that electoral competition encourages legislators to look for legislative opportunities that will increase their net political support. However, these opportunities are limited by a number of transaction costs: (1) the cost of legislative decision making and private participation, (2) the commitment problem of the legislature enacting a policy to ensure that future legislatures fund and maintain the policy, (3) agency costs (the costs incurred to induce administrative agencies to implement what was intended in the legislation and the losses legislators and other principals sustain by being unable to do so perfectly), and (4) the costs of risk and uncertainty.

Thus, in order to serve their interests well and protect their interests from being encroached by the future legislatures and the administrative agencies, the enacting legislature will adopt different institutional designs to solve the different transaction problems they meet. Horn argues, "[t]he legislators who are most likely to remain in power are those who are most successful in overcoming these transaction problems" (p. 14). He argues that the key to understanding these institutional designs is to focus on the relationship between legislators and their constituents and to recognize that constituents will exercise intelligent foresight. Moreover, in this relationship, the commitment problem is particularly acute because, "[s]overeignty is a distinguishing feature of government and implies that the incumbent legislature is unable to commit future legislatures. This creates political uncertainty that is at the heart of all political transactions" (p. 183).

Horn then applies the transaction cost model of public administration to three common types of public sector administrative activities: regulation, tax-financed production by bureaus, and sales-financed production by state-owned enterprises. In these cases, the transaction cost framework is used to examine choices made about the scope of authority delegated to the administrative level, the choice of regulatory agents, and administrative procedures. He argues that the enacting legislature facing agency costs must balance between these and other transaction costs. This then explains "why the legislature constrains 'its' ability to control its administrative agents" (p. 78).

Horn then uses this model to explain the institutional choice the enacting legislature makes when it turns to tax financed bureaus and examines the nature of expenditure controls, mandated expenditures, patronage, and civil service rules that constrain legislative influence over bureaucrats. Horn argues that two common features of government budgets in industrialized countries – mandated expenditures and expenditure controls – can be explained "in terms of the enacting legislature's need to overcome its commitment problem and, to a lesser extent, to control the agency costs" (p. 83). Again, commitment problems play an important role in explaining the design of budgetary institutions. This framework is also used to examine the existence of state-owned enterprises. In sum, in Horn's model, institutional choices by legislatures facing many different types of transaction problems set the context of the government budget, which in turn establishes the patterns of interaction among budgetary actors.

It should be noted that although Horn disagrees that his model intends to justify current institutional arrangements, his conclusion is that the modern institutions of administrative government are efficient for the purpose of serving the interests of the legislatures. Recently, Williamson (1999) has applied transaction cost theory to public bureaucracy and attempted to answer why certain types of transactions (so-called "sovereign transactions," such as foreign affairs) are only supplied by governments. After analyzing asset specificity, probity, and operating costs related to these transactions, Williamson concludes that, in this case, the institution of public bureaucracy can improve efficiency. Therefore, the transaction cost approach does not necessarily undermine the legitimacy of public institutions, as argued by some (Terry, 1998).

A Transaction Cost Model of Politics

In examining the policy-making process, Dixit (1996) developed the approach of transaction cost politics. Transaction cost politics (TCP) focuses on studying political processes from the transaction cost perspective. The central thrust in Dixit's theory is that "[m]any features and outcomes of the policy process can be better understood and related to each other by thinking of them as the results of various transaction cost and of the strategies of participants to cope with these costs" (pp. xiv–xv). He identifies three main forms of transaction costs: information impactedness, opportunism (moral hazard), and asset specificity.

Dixit argues that information impactedness is generally more severe in TCP than in transaction cost economics. Regarding opportunism, Dixit argues that agency relationships in TCP are much more complicated. There are frequent occurrences of multiple agency and common agency problems which make the transactions more complicated. As to asset specificity in TCP, Dixit discusses issues such as irreversible investment made by political parties, politicians with specific assets (in location, industries, etc.), pivotal political supporters with economic-specific investments, and cases where government is allowed to lock in a policy by a long-term commitment. To him, "political processes seem likely to be even more beset by transaction costs than are economic relationships" (p. 47). Dixit argues that in contrast to transaction cost economics, where market competition can select out inefficient governance structures, TCP has weaker and slower forces of evolutionary selection, exacerbated by the greater ability of political authorities to manipulate information which can slow down the forces of selection and evolution.

The central theme of Dixit's approach is that in the political arena in order to solve the problem of transaction cost, some economically inefficient governance structures will persist. Political governance structures that may appear inefficient may be a reasonable way of coping with the transaction costs. As Sayer indicates, this suggests that transaction cost constraints are as important, "just as much as we respect resource and technology constraints" (1999, p. 218).

Dixit then uses TCP to examine two cases: tax and expenditure controls in the U.S. and the General Agreement on Tariffs and Trade (GATT). His approach is to identify the transaction costs existing in these systems, the problems that they create, and to search for the mechanisms that have been developed for coping with the transaction costs. For tax and expenditure controls, the central question asked by Dixit is: why does the political process so persistently fail to achieve the goal of budgetary balance even after so many tax and expenditure reforms? He examines the constitutional rules of fiscal decisions and the resulting policies. He argues that constitutions are incomplete contracts and that procedures have particularly important effects on the policy outcome. The development of the committee structure in Congress is a good example of this. Before 1877, the House Ways and Means Committee was the central Congressional committee for fiscal affairs. Since then the number of committees involved has multiplied. Under this multiple committee system, each of these committees has its own priorities, "which it can better meet by raiding the overall budget. This is 'common resource' problem . . .The outcome is a Prisoners' Dilemma" (p. 120).

What Dixit attempts to demonstrate is that the true underlying political issues, "of which the procedures are merely symptoms" (p. 121) can be explored from the perspective of transaction cost. Dixit cites Penner's (1992) statement that, "budget rules that conflict too severely with political incentives cannot long endure" (p. 19). The institutional arrangements causing an apparent lack of discipline over expenditures is actually an underlying political equilibrium where those who benefit from these expenditures are powerful enough to preserve them. This viewpoint would imply that the dispersal of spending authority under the committee structure, which appears to be socially inefficient, helps reduce the political transaction costs faced by Congress. North's perspective further emphasizes this argument. North (1994) argues that "[i]nstitutions are not necessarily or even usually created to be socially efficient; rather they, or at least the formal rules, are created to serve the interests of those with the bargaining power to create new rules" (pp. 360–361).

Dixit's argument is related to Weingast and Marshall's (1988) research, which presents a theory of legislative institutions that parallels the theory of contractual institutions. They argue that like market institutions, legislative institutions reflect two key components: the goals or preferences of representatives and the relevant transaction costs. They theorize that legislative institutions enforce bargains among legislators, and that because of the form of bargaining found in legislatures, specific forms of non-market exchange often prove superior to market exchange when taking into account transaction costs. To them, legislative institutions like committee systems economize on transaction cost in the legislative process. Thus the concept of transaction costs is broadened to include decision-making costs over the life of the policy, as well as economic transaction costs.

A Transaction Cost Model of Budgetary Institutions

Viewing the budget as a contract, Patashnik (1996) argues that transaction cost theory can be used to examine the design of budgeting institutions. He states that although it is a mistake to import transaction cost theory into budgeting research without subjecting its premises to critical scrutiny, three claims can be made about budgeting:

(1) the costs of negotiating and enforcing budget contracts shape the budgetary process, and through it, the budgetary outcome;
(2) political actors deliberately craft institutional safeguards to add durability to their commitments; and
(3) budget reforms are unlikely to succeed if they fail to take into account both the potential for opportunistic political behavior and the inherent need of complex transactions for contractual safeguards (p. 191).

This model focuses on the cost of negotiating and enforcing the agreements by which policymakers allocate the government's resources. Patashnik discusses the behavioral assumptions of bounded rationality and opportunism, and compares them with the assumptions of competing budgetary models, then examines whether commitment costs and agency costs have structured budget actors' institutional choices. He indicates that in presuming omniscience and benevolence, the synoptic rationality models of budgeting are far too utopian to describe how budgeters actually behave. The transaction cost model is different from the synoptic rationality model. Transaction cost theory shares the assumption of bounded rationality with incrementalism. However, "it strongly rejects incrementalism's faith in the willingness of actors to moderate their demands or abide by norms of procedural fairness" (p. 192). Rather, transaction cost theory's assumption of opportunism enables it to explain many opportunistic behaviors in budgeting. It also broadens classical incremental theory beyond the environment and relations described by Wildavsky (1964). While Wildavsky described a process of consensual bargaining in a stable environment by a small community with enforcing norms, in transaction cost theory all of these characteristics are seen as variables that can change. Indeed, the changes in these characteristics might explain the transformation of U.S. federal budgeting.

In order to explore whether budgeters minimize transaction costs, Patashnik identifies the defining attributes of budget transactions and describes the performance features of alternative budgetary governance structures. He argues that certain kinds of budget transactions pose more severe transaction problems. The attributes that are especially important for budget contracts are political uncertainty, asymmetric information, and asset specificity. In the public sector, "high asset specificity characterizes transactions in which citizens make major life decisions – such as how much to save and when to retire – in the expectation of promised benefits" (p. 198). Social Security is an example of a program with high asset specificity. It has both a high level of institutional protection and is a program which elected officials have a strong incentive to protect. This is the strongest type of contract in the public sector.

Patashnik identifies three key performance features for budgetary governance: decision-making efficiency, flexibility, and credibility. Then, he uses these criteria to examine three types of governance structures: zero-base budgeting, annual appropriation budgeting, and entitlement budgeting. He writes that "transactions differ in their governance needs while governance structures vary in their performance attributes" (p. 200). In the private sector the hypothesis of discriminating alignment asserts that the efficient governance structure will emerge over the long run. However, this assertion is questionable in the public sector, as the competitive forces to stimulate this alignment are much weaker, and efficiency is often not the primary goal.

His investigation of the use of budget instruments – entitlement, multi-year appropriations, and tax expenditures – suggests that Congress has been more discriminating in its institutional choices than is commonly supposed. He concludes that,

> [T]ransaction cost theory appears to have the capacity to integrate a great deal of empirical knowledge about the design of contemporary budgeting institutions and to make contact with best insights from the traditional budgetary literature. . . . At a time when budgeting lacks a clear research agenda, transaction cost theory has the potential to offer a useful framework for organizing scholarly work. . . . The real question is not whether transaction costs matter for budgeting, but how transaction costs interact with other factors to produce the organizational phenomena that we seek to understand" (pp. 204–205).

A Transaction Cost Model of Budgetary Execution

Fred Thompson and L. R. Jones (1986) and Thompson (1993) have applied the concepts of transaction cost theory to policy implementation in general, and budget execution more specifically. To study administrative control in government contracting, they focus on two variables: the cost structure of production (increasing or decreasing returns to scale) and the heterogeneity of the outputs. This creates four typologies, which then helps determine whether responsibility for administration of the appropriation, grant, or contract should be placed with either individuals or organizations, and whether control should be *ex ante* or *ex post*. The optimal management control system designs associated with these four typologies are: outlay budgets, responsibility budgets, fixed-price contracts, and flexible-price contracts. They argue that managers should then use these typologies in deciding what controls are feasible and desirable. No single design is best in all situations; it depends on the cost structure and output heterogeneity variables. They go on to suggest that substantial cost savings are possible by matching the proper control designs to each public good or service. For example in naval shipbuilding, when management controls are appropriately matched to goods, control costs were a third less than the control cost when all components are produced internally, and half of what they would have been if all components were outsourced (Thompson, 1993). Thus, appropriate institutional design is necessary for efficient and effective budget execution.

Thompson and Jones apply this conceptual framework to the relationship between the budget "controller" and the agency in the design, execution, and monitoring of budgetary contracts. Their framework is a very helpful one in the development of this model as applied to budgeting. Conceptually, its only apparent shortcoming is the need to expand it to other phases of the budgetary process (formulation and adoption) and to other relationships (principally those involving legislators). Empirically, like most of these theories, it needs to be tested. Normatively, it is subject to criticism for being hierarchical.

Other Transaction Cost Models of Fiscal Policy

Tax Policy

Some research from this perspective has been applied to issues of taxation. Williamson (1985) wrote that public finance is an area where transaction cost economics "has made only limited contributions but holds out considerable promise" (p. 394). Since then, application of transaction cost economics to public finance has increased.

First, the transaction cost concept is beginning to be integrated into tax theory. Riggall (1997) examined transaction costs as one of the elements to consider in the design of a comprehensive optimal tax. Hauler (1998) looked at the effect of transaction costs on cross-border shopping. Eggert and Haufler (1998) incorporated transaction costs into an international capital tax model.

Second, transaction cost theory has also been applied to the study of subsidies. Wolfson (1990) argues that traditional tax analysis is not adequate for creating a conceptual framework for a theory of subsidization, and that optimal taxation theory has not yet progressed to the point of being useful for policy analysis. Wolfson argues that subsidized activities should minimize information and transaction cost.

Third, transaction cost theory has also been applied to "club theory" originated by Buchanan (1965), Tiebout (1956), Wiseman (1957), and Olson (1965). Those models assumed away any costs in administration and monitoring programs. In the 1990s, the effects of exclusion, monitoring, and enforcement costs on the institutional structures of the clubs have been examined by Helsley and Strange (1991), Lee (1991), and Silva and Kahn (1993). For example, Helsley and Strange examined the competitive provision of a club good in the presence of exclusion costs. With no exclusion costs, exclusion is not optimal, however exclusion may improve welfare when these costs are considered. Lee extends this analysis by introducing imperfect information in the choice of the degree of exclusion, where the club cannot distinguish between high and low demanders.

Expenditure Policy

Transaction cost theory has also been applied to the issue of public goods provision. Ferris and Graddy (1991) address the practice of contracting as an alternative to public supply of services. In their discussion, contractor choice is modeled as one in which the decision-maker minimizes service delivery costs, both production and transaction costs, subject to political and fiscal constraints. Clingermayer and Feiock (1997) apply a similar model, incorporating the influence of leadership turnover into this decision.

Weimer and Vining (1996) use transaction cost theory to critique the expenditure control budget advanced by Osborne and Gaebler (1992) which allows managers to carry over unspent funds from one fiscal period to the next. Weimer and Vining state that transaction cost theory, "suggests that we should be skeptical about the expenditure control budget's use in circumstances in which the manager and the central budget office do not already have a trusting relationship" (p. 105). Similarly, Christensen (1992) examines top-down budgetary control efforts in Denmark and finds that they are unsuccessful because they neglect to change the incentives facing actors at the micro-level. Political actors have incentives to break the budgetary rules they previously set, undermining the credibility of the reforms. The micro-level perspective he advocates incorporates the assumptions of transaction cost theory.

Frant (1996) uses transaction cost economics to examine the issues of earmarking, public authorities, and civil service personnel systems. Smith and Bertozzi (1998) test a principal-agent model of budgeting applied to New York State. They conclude that it is a more accurate model of micro-level budgeting than either incrementalism or Rubin's decision-making clusters model.

Finally, Zerbe and McCurdy (1999) argue that the concept of market failure that has been used to justify government intervention in the market economy is flawed and that the transaction cost concept is a better one in evaluating the normative case for intervention. The Coase theorem made it clear that externalities and other market failures exist only because of transaction cost impediments. Further, the efficient response to any market failure is not necessarily government intervention, but rather a search for the institutional arrangement that minimizes the production and transaction costs. They argue for an empirical basis for intervention and write that there are few general rules for efficient solutions to market inefficiencies. For example, cultural norms of trust and honesty can substitute for government action but these vary from one government to another. Zerbe and McCurdy's argument is a normative one focused on policy at a broad level, which demonstrates the centrality of institutions to public intervention strategies. Their argument that the correct solution is dependent on the actual design of alternative institutions is persuasive and significant.

Assessment

This literature review shows that some theorists have recognized transaction cost theory as a potentially productive theoretical framework to understand budgetary activities, processes, and institutions. However, no one has presented a unified and complete theory of budgeting from the perspective of transaction cost theory. Even in the work of Thompson, Horn, and Dixit, the discussion of budgetary issues is a by-product of a transaction cost model of public administration and policy-making.

Patashnik's work is the only one focused on budgeting from the perspective of transaction cost theory. However, many fundamental issues for a transaction cost budgetary theory are yet to be explored. The relative utility of transaction cost theories as positive versus normative models of budgeting is unclear. Finally, the limitations of a transaction cost budgetary theory have not been well discussed.

ADAPTING THE THEORY TO THE PUBLIC SECTOR

To begin to build a transaction cost budget theory, the framework of transaction cost politics is a more suitable approach than transaction cost economics.

Transaction Cost Politics

The term of transaction cost politics originated in North's article *A Transaction Cost Theory of Politics* (1990b), and Wilson (1989) briefly mentions it as an idea that is unexplored but worth attention. Moe (1990a) questioned whether agency theory or transaction cost economics could carry over from the private to the public sector. Moe identified some important differences between the economics and politics of organizations, including: (1) the basic comparison between markets and hierarchies in private sector transactions does not apply for most of the goods and services supplied by government; (2) political actors may not be able to structure their relationships efficiently because they are unable to sell their rights to exercise public authority, and also the need for compromise in politics give rise to expediency rather than efficiency in the design of public bureaus; (3) public authority is involuntary; and (4) asset specificity cannot be applied to public sector (Moe, 1990a). Moe questioned whether either transaction costs in general or asset specificity in particular is the key to choice in the world of politics.

In contrast to Moe, North is less concerned with the public bureaucracy as an organizational form than he is with the efficiency of the public choice process.

To North, the efficiency of politics is to be judged by examining how closely a political market can "approximate a zero transaction cost result" (1990b, p. 360). He examines the source of high transaction costs in political markets: differing subjective models of decision-makers, and measurement and enforcement costs of agreements. He concludes that political markets are more beset by transaction costs, and thus more prone to inefficiency. Dixit (1996) develops a broader theme of transaction cost politics, and comes to the same conclusion as North. When North and Dixit recognize that the political process is beset by transaction costs, they are right. But they are not right when they claim that this demonstrates that political markets are inefficient. The main feature of TCP is that sometimes efficient political institutions *increase* transaction costs in political markets.

Should Government Institutions Minimize Transaction Cost?

Theories of transaction cost economics differ, such as the approaches of North and Williamson. In North's model, transaction cost is the dependent variable while institutional arrangements are independent variables. In Williamson's model, transaction cost is an independent variable while the governance structure is the dependent variable. Further, North considers human nature complex while Williamson uses the assumption of opportunism. However, both concur that efficient governance structures are those that would minimize transaction costs. Minimizing transaction costs lies at the core of normative transaction cost economics.

However, when this logic is applied to examine political institutions in general, and budgetary institutions in particular, we face a puzzling fact that apparently some institutions are deliberately designed to *increase* not decrease transaction costs. Bicameralism, separation of powers, and federalism are all institutions that increase transaction costs. In their explanation of the evolution of political institutions in 17th century England, North and Weingast (1989) conclude that the separation of power as a basic political institution limits governmental intervention to tax or confiscate private property and forces government to commit to honor its agreements. They state that, "increasing the number of veto players implied a larger set of constituencies could protect themselves against political assault, thus markedly reducing the circumstances under which opportunistic behavior by the government could take place" (p. 162). However, North and Weingast do not mention that increasing the number of veto players in the fiscal and political institutions almost inevitably increases the transaction costs. Failing to recognize that in the political arena it is sometimes desirable to increase transaction costs to protect the public interest leads to the conclusion that the political market is inefficient because of its high transaction costs. This focus on efficiency to the neglect of other criteria, such as liberty and justice, leads to a premature condemnation of political institutions.

Vira (1997) concurs, arguing that, "neoclassic institutionalism pays excessive attention to the lowering of political transaction costs as a strategy to facilitate socially desirable institutional change" (p. 773). He argues that, "it is debatable whether lower transaction costs are always socially desirable," (p. 772) and that high transaction costs may be desirable to prevent powerful groups from appropriating benefits in an unjust or unfair way. Laffont (1999) also argues that the increase in transaction costs caused by the separation of powers is helpful to prevent the capture of government by interest groups.

One approach to this problem is to think of costs broadly and in the long run. North (1990a), among others, defines total costs as the sum of transaction costs and "transformation," or production costs. Any external costs also need to be considered to achieve socially efficient production. Institutional features that increase transaction costs may lower other costs enough to be more efficient than other arrangements with low transaction costs but high production costs or external costs. To judge any institutional feature by the efficiency criterion, it is necessary to consider the effects of institutional attributes on all associated costs. However, this might expand the concept of cost too broadly. Alternatively, the theory could simply realize that efficiency is just one goal for society, and that liberty, justice, and equality are other goals that cannot meaningfully be calculated. Desirable institutions attempt to serve all social goals as well as possible.

If nothing else, a transaction cost theory of budgeting would systematically focus on institutional features as both independent and dependent variables. We should study institutional design to understand the effects a given institution has on the budgetary process and, in turn, on governance structure rather than focusing on its more narrow impact on political or economic transaction costs.

FOUNDATIONS FOR A TRANSACTION COST BUDGETING THEORY

In order to build a transaction cost theory of public budgeting, the following issues are fundamental: Is a budget a contract? Can budgetary decisions be viewed as transactions? What are the main behavioral attributes of the budgetary actors, and what are the main environmental attributes of the budgetary transactions? And, what is the relationship between transaction costs and governance structure?

Viewing Budgets as Contracts

As Patashnik (1996) argues, the concept of the budget as a contract is not new although its theoretical power has not been fully exploited. Wildavsky (1964) wrote that "[v]iewed in another light, a budget may be regarded as a contract" (p. 2). Viewing a budget as a contract immediately raises the crucial issue of enforceability, as addressed by the discussions of uncertainty by Horn (1995) and Moe (1990b). Transaction cost theory is useful for studying contractual relations. As Williamson (1985) argues, "[v]irtually any relation, economic or otherwise, that takes the form of or can be described as a contracting problem, can be evaluated to advantage in transaction cost economic terms" (p. 387). This generality is one of the important virtues of this theory.

Behavioral Attributes of Actors

From the vantage point of transaction cost economics, limitations in the rationality of decision-makers and their inclination to be opportunistic imply that under conditions of uncertainty, information impactedness, smaller number exchange, and asset specificity transactions become expensive and hazardous (Williamson, 1975, 1985). Therefore, in order to develop the concept of budgetary transaction costs, it is necessary to examine the behavioral attributes of budgetary actors and the environmental attributes of budgetary transactions. This section examines three attributes of budgetary actors: bounded rationality, opportunism, and risk preference. The next section examines the attributes of budgetary transactions.

Bounded Rationality

Transaction cost economics deviates from mainstream economics by assuming bounded, rather than synoptic, rationality. Bounded rationality has been used in many models of budgeting, such as incrementalism and organizational process models. Decision-making under incrementalism's version of bounded rationality is restricted. A narrow range of information is used; typically the base budget is not reconsidered and the focus is on the increment. Bounded rationality under the transaction cost perspective is broader. Actors use as wide a range of sources of information as possible, and attempt to reduce uncertainty and to broaden their exchange opportunities. Further, the transaction cost perspective argues that bounded rationality alone cannot explain the budgeting process and outcomes. Here incrementalism and a transaction cost approach differ, because as Patashnik writes, incrementalism has a strong faith in "the willingness of actors to moderate their demands or abide by norms of procedural fairness" (p. 192), while the transaction cost approach assumes opportunistic behavior.

Opportunism

Is opportunism a valid assumption about the behavior of budgetary actors? Opportunism may or may not exist in any situation, but this approach incorporates it as a variable, allowing it to vary from absent to rampant. It seems reasonable to believe that political actors in a negotiation may consider opportunistic behavior and its consequences. Assuming it away is naïve. Opportunistic behavior can be seen *ex ante* and *ex post* of the contract, as Thompson and Jones write. In the process of negotiating budgetary contracts, some actors may take advantage of information asymmetry, and induce other actors to agree to a budgetary contract that is against their interests. Moreover, even if efficient and just contracts have been reached, the commitment problem remains. Will all the budgetary actors commit to the budgetary contract they signed?

Incrementalism assumes that actors live by their agreements, and that any problems in budget execution can be corrected in the next budget cycle. In contrast, transaction cost theory argues that opportunism may exist in budgetary activities. Two main forms of opportunism are identified: *legislator opportunism*, and *bureaucratic opportunism*. The former has been explored by Horn and Shepsle (1989), Shepsle (1992), Macey (1992), and Horn (1995) under the term of "coalition drift" or "legislative drift" which is the contracting problem facing enacting politicians. This type of opportunism has two forms: (1) opportunism between elected officials and taxpayers where elected officials may promise something to the taxpayers but in reality may not deliver and (2) opportunism among the elected officials. This mainly arises from uncertainty over political property rights (Moe, 1990b). Bureaucratic opportunism has been explored by the "McNollgast" model under the term of "bureaucratic drift" (McCubbins, Noll & Weingast, 1987, 1989), and in the case of budgeting explored by Niskanen (1971).

Risk Preference

Traditional transaction cost economics relies on three behavioral assumptions: bounded rationality, opportunism, and risk neutrality. However, the assumption of risk neutrality greatly limits the explanatory strength of the transaction cost theory. Chiles and McMackin (1996) developed a transaction cost model which integrates variable risk preferences into Williamson's (1991) model. In their model, effects of trust on the governance structure can be explained within the framework of transaction cost theory. Exploring the risk preference of budgetary actors is also important for a transaction cost budgeting theory because this preference will affect actors' selections of contracts and institutions. If most actors are risk-averse, a low level of asset specificity, political uncertainty, or information asymmetry will be acceptable. If most budgetary actors are risk-seeking, high levels of these features will be preferred. Also, trust among actors matters, because trust will affect their perceived risk, and in turn the selections of budgetary contracts and governance structures. If budgetary actors trust each other, the need for exchanging information is reduced, transaction costs are saved, and safeguard mechanisms are simple or unnecessary for many budgeting activities.

In sum, budgetary contracts and governance structures will be affected by risk preference and trust. For example, some have suggested that agencies should be allowed to transfer unspent funds from one period to the next. However, if the agency managers and the central budget office do not already have a trusting relationship, this may not work because, "the manager must anticipate the possibility that the budget office will expropriate the savings by reducing the base budget in the next period" (Weimer & Vining, 1996, p. 105). In this case, rules may be required to make a credible commitment not to expropriate the savings. However if there is a trusting relationship, often no new rules are necessary.

Attributes of Budgetary Transactions

Transaction cost theory argues that the attributes of actors alone cannot explain many transaction problems in the real world; only in some environments can opportunistic actors succeed. These physic dimensions of transactions include: uncertainty, information distribution, asset specificity, and small number exchange. Transaction cost economics maintains that in the presence of these problems, it is easy for actors to behave opportunistically. Transaction problems may arise and require governance structures or safeguard mechanisms.

Uncertainty

Horn (1995) writes that many budgetary institutions are designed by legislatures to solve the transaction problem resulting from the political uncertainty. Patashnik (1996) points out that "[e]lected officials are free to enter into the budget contracts of their own choosing, but these contracts are vulnerable to modification or even nullification if future politicians have opposing policy interests" (p. 194). Anticipating this type of opportunism, elected officials may seek to design and implement safeguard mechanisms that shape budgetary institutions. For example, the financing of social security benefits from the payroll taxes of current workers created a system particularly difficult to dismantle, thus solving the commitment problem for the enacting Congress.

Interest groups also face the problem of political uncertainty, and will seek to solve this political uncertainty through "structure choice" (Moe, 1990a). They will seek to protect themselves from political uncertainties in the budgetary process, which will affect the strategies and the budgetary governance structures they will support.

Information Distribution

Information costs are a major part of the costs of making and enforcing budgetary contracts. These costs will be affected by the distribution of information among budgetary actors which in turn will affect their strategies. Information asymmetry may increase information cost for certain actors, and make them vulnerable to the opportunistic behavior of other actors.

The information distribution between the legislature and the bureaus may be the most controversial. Niskanen's (1971) budget-maximizing model is built on specific assumptions about information asymmetry between bureaus and the legislature. However, recent models, especially Dunleavy (1991) and Waterman and Meier (1998), have a more realistic range of assumptions about this information distribution. Information distribution among actors is a variable, not a constant and should be modeled as such. Bureaucrats, or any other actors, may not be able to behave opportunistically even if they want to.

Asset Specificity

While appreciating Moe's suspicion of applying asset specificity to politics, we argue that asset specificity exists in budgetary transactions. As Schick (1988) indicates, budgeting has two basic elements: claiming and allocating resources. For multi-year expenditures such as public investments, after incurring the first year expenditure, these funds become "program-specific" expenditures. In this case, the advocates of this program may use this asset specificity to support their new claims for future allocations.

Based on the work of Dixit, we examined some other budgetary asset specificities. We have identified five types of budgetary asset specificities, recognizing that this is not exhaustive. They are as follows:

(1) Irreversible investment of a political party. If a political party has invested more in a certain policy, and their reputation and future benefits are closely related to that policy, it is reasonable to argue that this political party will be a strong supporter of that policy.
(2) Irreversible economic investment of interest groups. If an interest group's investment is sunk in a certain activity or industry, it may seek to lobby government to offer protection for that industry.

(3) Asset specificity of elected officials. This refers to the specific locations or industries from which the elected officials receive political support. This type of asset specificity may affect elected officials' budgetary activities. In order to continue to win political support from their specific locations or industries, elected officials will seek to win budgetary expenditures for their specific constituents. How will this kind of asset specificity affect the budgetary strategies of elected officials, and the budgetary contracts and governance structures they will support? If all elected officials have this type of asset specificity, what is the best budgetary governance structure? Will the budgetary governance structure be the norm of "univeralism" or "constrained universalism" (Inman & Fitts, 1990)?

(4) Long-term commitment. Long-term commitments made by government will lock the government into certain kinds of policies and programs, which then commits future budgets.

(5) Program and organization-specific human capital. Human capital may be specific to certain types of jobs and organizations. If the human capital of bureaucrats are organization-specific, they will be more likely to defend the mission and budget of their organization.

This suggests that asset specificity exists in budgeting, and will affect the transaction cost and, in turn, the selection of budgetary contracts and governance structures. However, all of these are still untested hypotheses. Future research and empirical tests are required.

Transaction Cost and Governance Structure

The core of a transaction cost theory of budgeting is the relationship between transaction cost and budgetary governance structure. There are two general lines of exploration. The first, developed by North, focuses on examining how institutional arrangements affect transaction cost. This is positive research. The second, developed by Williamson, seeks to find the most efficient governance structure to solve the transaction cost problem. This is normative research. We argue for applying both of these approaches to budgetary theory.

Positive Theory

As discussed above, the attributes of budgetary actors and transactions are critical variables affecting the transaction cost of activities, and in turn, budgetary institutions. For example, if trust has been developed among budgetary actors, they will not be inclined to behave opportunistically, lowering transaction costs and allowing for a less rule-bound governance structure. Transaction costs in budgetary transactions are composed of: (1) *ex ante* costs of contracting, such as costs of drafting, negotiation, and safeguarding a budgetary contract, and (2) *ex post* costs, such as measurement costs, monitoring costs, and enforcement costs of the budgetary contract.

A positive transaction cost theory of budgeting would then apply these attribute variables to explain the characteristics of institutional design. For example, in examining the practice of outsourcing by governments, one would hypothesize that these factors would affect the total cost (production cost plus transaction cost) of the alternatives, and therefore the outcome. Identifying variation in these variables and testing their influence on the outcome would be a test of this theory.

The potential of this approach as a positive model depends on the effects of transaction costs, or any other costs, on the decision-making of government officials. With the assumption of competition in the private sector, it is reasonable to predict that over the long run firms that do not economize on total costs will go out of business. However, this is not always a safe assumption in the public sector for two reasons. First, some governments do not face serious competition, and second, because cost minimization is not the only goal of government.

Normative Theory

As Williamson (1998) argues, much of the predictive content of transaction cost economics works through the discriminating alignment hypothesis: "transactions, which differ in their attributes, are aligned with governance structures, which differ in their cost and competence, so as to effect a (mainly) transaction-cost economizing result" (p. 75). Analyses of the attributes of actors and budgetary transactions show that budgetary contracts are easily beset by contractual hazards, which may make desirable transactions unlikely to occur, while making undesirable transactions more likely. Therefore, the actor who perceives a contractual hazard will seek to "embed transactions in governance structures that have hazard-mitigating purpose and effect" (p. 76). Other attributes that deserve attention are the cost of the governance structure and the legitimacy of the governance structure.

While examining governance structures, we face a difficult task because, on one hand, transactions vary in both their attributes and their governance requirements, while governance structures vary in their performance attributes and their governance cost. Hence, the challenge for reformers is to match governance structures to budgetary transactions. This is essentially a process of institution innovation. For any reform proposal, the incentives for supply of the proposed new governance structure and the effect of the governance structure on transaction costs need to be carefully examined in determining its feasibility.

APPLICATIONS

There are a number of areas of research in public budgeting to which a transaction cost model could be productively applied. This research can be either normative or positive. The central value of the normative perspective is efficiency. Of course some decision-makers or researchers may not share this value, so that is a limitation. Similarly, positive research would attempt to understand what happened using the efficiency lens. To the degree that this was not the goal of the government under study, the theory would also be limited.

Despite these shortcomings, in certain situations and for certain government activities, efficiency is an appropriate criterion. For example, in many of the areas of financial management, it is not much of a leap of faith to assert that governments attempt to achieve their objectives in the most efficient way (including transaction costs), and not too bold to assert that they should (Van Reeth, 2000). Managing cash, debt, investments, pensions, risk, and purchasing are "housekeeping" functions of government that most citizens and administrators want to do as efficiently as possible. Of course, certain elected or appointed officials may have other goals, such as doing business with local as opposed to outside contractors, investing in politically popular funds, outright nepotism, and so on. However, efficient and neutrally competent administration has been one of the central values in U.S. public administration, so the goal of efficiency here is not at odds with the prevailing normative value. If in doing positive research on these topics, researchers can identify deviations from efficient practice and then examine the values inherent in that deviation, they can then make a normative judgment about the practice.

So for example, governments face different transaction costs in purchasing. A city government in a large metropolitan area has many potential suppliers, and can force them to compete on terms most beneficial to the city. In comparison, a small government in an isolated rural area has many fewer potential suppliers in the area and, therefore, might be likely to pursue means other than individual market purchases, such as cooperative purchasing pools or long-term negotiated agreements. A cross-sectional analysis of this practice would be a good application of this theory. Further, the advent of on-line purchasing presumably has created an important opportunity for these governments to save on both their pecuniary costs and their non-pecuniary costs of purchasing. Looking at changes in purchasing practice over time in response to this environmental influence would therefore be a test of transaction cost theory.

As mentioned above, a similar question that can be examined from this perspective is the decision to out-source the production and delivery of public services. In either of these cases, one could examine either the normative question – Is the government managing in the most efficient way? – or the positive question – What independent variables have changed that should induce a change in the dependent variable (here, purchasing or outsourcing practice, or more broadly, governance form)?

Another fruitful area for future research from this perspective would be studying appropriation rules. As Patashnik has pointed out, there are a variety of ways that funds are allocated under even the simplest public budgeting systems. Most governments use fund accounting, which segregates funds according to specific rules, so that certain revenues fund certain expenditures. This institutional feature is a rule that governs transactions, and it is appropriate to analyze this practice from either a positive or normative transaction cost perspective. For example, in most situations the creation of a separate debt service fund would be expected to minimize long-term social costs, as the lower interest rates attainable and the reduced risk associated with this practice would outweigh the *ex ante* contracting costs, and the *ex post* costs of administering the fund and enforcement. The same is generally expected to be true for capital project funds. If these conditions hold, then the normative perspective would recommend these practices. Positive research could examine the evolution of such practices as well as cross-sectional variation in their use, and the effects of these governance structures on institutions and policies.

More complex appropriation rules – such as entitlements, trust funds, earmarking, impoundment rules, and more specific rules such as those that were adopted as a part of the Gramm-Rudman-Hollings Act, or the 1990 Budget Enforcement Act – can also be analyzed in the same way. For example, trust funds are heavily used to fund transportation. From a positive perspective, case studies could explain the evolution of this sort of an arrangement; first the negotiation and adoption of the policy, then the implementation of the policy and its enforcement and maintenance over time. Perhaps the same sorts of questions could be analyzed using aggregate data sets. From a normative view, researchers could contrast the beneficial aspects of this sort of an arrangement (such as certainty and asset specificity) compared to the negative aspects (such as opportunism and reduced ability for comprehensive budget review). The historical perspective of transaction cost theory would be especially helpful here to describe the development of such a fund, and its impact on the institutional structure of the transportation agencies, their power relative to the legislature, and the congruence of this arrangement with the interest group and citizen demands.

Yet another popular area of study in budgeting that could be productively studied from this perspective are the various rationalist reforms of the budgeting process that were proposed and, to some degree, implemented in recent years. Aaron Wildavsky's (1969) famous comment that "no one knows how to do program budgeting" suggests that the high transaction costs of these reforms and their potential to upset existing balances of power based on information distribution and other budgetary transaction attributes might help explain their limited implementation and durability.

These are but a few examples that attempt to illustrate the possibilities for application of this theoretical perspective to mainstream practical topics within public budgeting and financial management. This theory has been applied to the empirical study of accounting (for a helpful review, see Van Reeth, 2000) but, as this chapter indicates, only to a very limited degree in public budgeting and finance. It is a ripe topic that could be the source of many future dissertations and research studies, and good research in this area would ultimately help us judge the merits of the theory.

CONCLUSION

Transaction cost theory is powerful and general, thus making it an attractive theory to adapt to the public sector. The concept of a transaction is general enough that most government fiscal behavior can fit within the general framework. Further, we know that concepts like risk, opportunism, uncertainty, information distribution, and asset specificity are often relevant in explaining budgetary exchanges. Capturing these concepts as variables that explain outcomes would be a step in the right direction for budgeting research. However the model needs to incorporate more political concepts to truly apply to public budgeting. The focus on efficiency as the main, if not only, value is limiting. Also, the discriminatory alignment hypothesis is probably of limited relevance to positive models of government. Further, we need to better judge institutional features that increase budgetary transaction costs. Finally, the abstract nature of the theory may make it difficult to operationalize the relevant variables.

With further adaptation, this theory has the potential to serve as an organizing rubric to better compare the findings of disparate studies. Budgeting research needs to be able to turn contextual descriptions into variables, which can allow for better generalization. The historical and comparative potential of transaction cost theory, as articulated by North, is very attractive. Other theories of budgeting have many merits, and it is unclear at this time if this model is preferable. But it is clear to us that it does deserve further investigation.

REFERENCES

Bryson, J. M., & Ring, P. S. (1990). A transaction-based approach to policy intervention. *Policy Science, 23*, 205–229.

Buchanan, J. M. (1965). An economic theory of clubs. *Economica, 32*, 1–14.

Butler, R. (1983). A transactional approach to organizing efficiency: Perspectives from markets, hierarchies, and collectives. *Administration & Society, 15*, 323–362.

Chiles, T. H., & McMackin, J. F. (1996). Integrating variable risk preferences, trust, and transaction cost economics. *Academy of Management Review, 21*, 73–99.

Christensen, J. G. (1992). Hierarchical and contractual approaches to budgetary reform. *Journal of Theoretical Politics, 4*, 67–91.

Clingermayer, J. C., & Feiock, R. C. (1997). Leadership turnover, transaction costs, and external city service delivery. *Public Administration Review, 57*, 231–239.

Dixit, A. K. (1996). *The making of economic policy*. Cambridge: The MIT Press.

Dunleavy, P. (1991). *Democracy, bureaucracy and public choice*. New York: Prentice Hall.

Eggert, W., & Haufler, A. (1998). When do small countries win tax wars? *Public Finance Review, 26*, 327–361.

Ferris, J. M., & Graddy, E. (1991). Production costs, transaction costs, and local government contractor choice. *Economic Inquiry, 29*, 541–555.

Frant, H. (1996). High-powered and low-powered incentives in the public sector. *Journal of Public Administration Research and Theory, 6*, 365–381.

Hauler, A. (1998). Asymmetric commodity tax competition. *Journal of Public Economics, 67*, 135–144.

Heckathorn, D. D., & Maser, S. M. (1987). Bargaining and the sources of transaction costs: The case of government regulation. *Journal of Law, Economics, and Organization, 3*, 69–98.

Helsley, R. W., & Strange, W. C. (1991). Exclusion and the theory of clubs. *Canadian Journal of Economics, 24*, 888–899.

Horn, M. J. (1995). *The political economy of public administration*. Cambridge: Cambridge University Press.

Horn, M. J., & Shepsle, K. A. (1989). Commentary on "Administrative arrangements and political control of agencies." *Virginia Law Review, 75*, 99–508.

Inman, R. P., & Fitts, M. A. (1990). Political institutions and fiscal policy: Evidence from the U.S. historical record. *Journal of Law, Economics, and Organization, 6* (Special Issue), 79–132.

Laffont, J.-J. (1999). Political economy, information and incentives. *European Economic Review, 43*, 649–669.

Lee, K. (1991). Transaction costs and equilibrium pricing of congested public goods with imperfect information. *Journal of Public Economics, 45*, 337–362.

Macey, J. R. (1992). Organization design and the political control of administrative agencies. *Journal of Law, Economics & Organization, 8*, 93–110.

Maser, S. M. (1986). Transaction costs in public administration. In: D. J. Calista (Ed.), *Bureaucratic and Governmental Reform* (pp. 55–71). Greenwich: JAI Press Inc.

McCubbins, M. D., Noll, R. G., & Weingast, B. R. (1987). Administrative procedures as instruments of political control. *Journal of Law, Economics & Organization, 3*, 243–277.

McCubbins, M. D., Noll, R. G., & Weingast, B. R. (1989). Structure and process, politics and policy: Arrangements and the political control of agencies. *Virginia Law Review, 75*, 431–482.

Moe, T. (1984). The new economics of organization. *American Journal of Political Science, 28*, 739–777.

Moe, T. (1989). The politics of bureaucratic structure. In: J. E. Chubb & P. E. Peterson (Eds), *Can the Government Govern?* (pp. 267–329). Washington: The Brookings Institution.

Moe, T. (1990a). The politics of structural choices: Toward a theory of public bureaucracy. In: O. E. Williamson (Ed.), *Organization Theory: From Chester Barnard to the Present and Beyond* (pp. 116–153). New York: Oxford University Press.

Moe, T. (1990b). Political institutions: The neglected side of the story. *Journal of Law, Economics, and Organization*, *6*, 213–253.

Niskanen, W. (1971). *Bureaucracy and representative government*. Chicago: Aldine Atherton.

North D. C. (1990a). *Institutions, institutional change and economic performance*. Cambridge: Cambridge University Press

North, D. C. (1990b). A transaction cost theory of politics. *Journal of Theoretical Politics*, *2*, 355–367.

North, D. C. (1994). Economic performance through time. *American Economic Review*, *84*(3), 359–368.

North, D. C., & Weingast, B. R. (1989). The evolution of institutions governing public choice in 17th century England. *Journal of Economic History*, *XLIX*, 803–832.

Olson, M. (1965). *The logic of collective action*. Cambridge: Harvard University Press.

Osborne, D., & Gaebler, T. (1992). *Reinventing government: How the entrepreneurial spirit is transforming the public sector*. Reading, Mass: Addison Wesley.

Patashnik, E. M. (1996). The contractual nature of budgeting: A transaction cost perspective on the design of budgeting institutions. *Policy Science*, *29*, 189–212.

Penner, R. (1992). Political economics of the 1990 budget agreement. In: M. H. Kosters (Ed.), *Fiscal Politics and the Budget Enforcement Act* (pp. 1–19) Washington, D.C.: AEI Press.

Pouder, R. W. (1996). Privatization services in local government: An empirical assessment of efficiency and institutional explanation. *Public Administration Quarterly*, *20*, 103–127.

Riggall, K. (1997). Comprehensive tax base theory, transaction costs, and economic efficiency: How to tax our way to efficiency. *Virginia Tax Review*, *17*, 295–340.

Sayer, S. (1999). New political economy. *Journal of Economic Surveys*, *13*, 211–224.

Silva, E. C. D., & Kahn, C.M. (1993). Exclusion and moral hazard: The case of identical demand. *Journal of Public Economics*, *52*, 217–235.

Schick, A. (1988). An inquiry into the possibility of budgetary theory. In: I. Rubin (Ed.), *New Directions in Budget Theory* (pp. 59–69). Albany: State University of New York Press.

Shepsle, K. A. (1992). Bureaucratic drift, coalitional drift, and time consistency: A comment on Macey. *Journal of Law, Economics & Organization*, *8*, 111–118.

Smith, R. W., & Bertozzi, M. (1998). Principals and agents: An explanatory model for public budgeting. *Journal of Public Budgeting, Accounting & Financial Management*, *10*, 325–353.

Terry, L. D. (1998). Administrative leadership, neo-managerialism, and the public management movement. *Public Administration Review*, *58*, 194–200.

Thompson, F. (1993). Matching responsibilities with tactics: Administrative controls and modern government. *Public Administration Review*, *53*, 303–318.

Thompson, F., & Jones, L. R. (1986). Controllership in the public sector. *Journal of Policy Analysis and Management*, *5*, 547–571.

Tiebout, C. M. (1956). A pure theory of local expenditure. *Journal of Political Economy*, *64*, 416–424.

Twight, C. (1994). Political transaction-cost manipulation: An integrating theory. *Journal of Theoretical Politics*, *6*, 189–216.

Van Reeth, W. (2000). The bearable lightness of being: An explorative research on the uneven implementation of performance-oriented budget reform across agencies. Unpublished doctoral dissertation, Public Management Institute, Catholic University of Leuven, Belgium, forthcoming.

Vining, A. R., & Weimer, D. L. (1990). Government supply and government production failure: A framework based on contestability. *Journal of Public Policy, 10*, 1–22.

Vira, B. (1997). The political Coase theorem: Identifying difference between neoclassic and critical institutionalism. *Journal of Economic Issues, XLI*, 761–779.

Waterman, R. W., & Meier, K. J. (1998). Principal-agent models: An expansion? *Journal of Public Administration Research and Theory, 8*, 173–202.

Weimer, D. L. (1992). Claiming races, broiler contracts, heresthetics, and habits: Ten concepts for policy design. *Policy Science, 25*, 135–159.

Weimer, D. L., & Vining, A. R. (1996). Economics. In: D. F. Kettl & H. B. Milward (Eds), *The State of Public Management* (pp. 92–117). Baltimore: The Johns Hopkins University Press.

Weingast, B. R. (1984). The congressional-bureaucratic system: A principal agent perspective (with applications to the SEC). *Public Choice, 44*, 147–191.

Weingast, B. R., & Marshall, W. J. (1988). The industrial organization of Congress; Or why legislatures, like firms, are not organized as markets. *Journal of Political Economy, 96*, 132–163.

Wildavsky, A. (1964). *The politics of the budgetary process*. Boston: Little Brown and Company.

Wildavsky, A. (1969). Rescuing policy analysis from PPBS. *Public Administration Review, XXIX*, 189–202.

Williamson, O. E. (1975). *Markets and hierarchies: Analysis and antitrust implications*. New York: Free Press.

Williamson, O. E. (1985). *The economic institutions of capitalism*. New York: Free Press.

Williamson, O. E. (1991). Comparative economic organization: The analysis of discrete structure alternatives. *Administrative Science Quarterly, 36*, 269–296.

Williamson, O. E. (1998). The institutions of governance. *American Economic Review, 88*(2), 75–79.

Williamson, O. E. (1999). Public and private bureaucracies: A transaction cost economics perspective. *Journal of Law, Economics, and Organization, 15*, 306–342.

Wilson, J. Q. (1989). *Bureaucracy: What government agencies do and why they do it*. New York: Basic Books, Inc.

Wiseman, J. (1957). The theory of public utility price – An empty box. *Oxford Economic Papers, 9*, 56–74.

Wolfson, D. J. (1990). Toward a theory of subsidisation. *De Economist, 138*, 107–123.

Zerbe, R. O., & McCurdy, H. E. (1999). The failure of market failure. *Journal of Policy Analysis and Management, 18*, 558–578.